T0362111

Combustion Ash Residue Management

An Engineering Perspective

Second Edition

Combustion Ash Residue Management
An Engineering Perspective

Second Edition

by

RICHARD W. GOODWIN, Ph.D., P.E.

Amsterdam • Boston • Heidelberg • London • New York • Oxford
Paris • San Diego • San Francisco • Singapore • Sydney • Tokyo
William Andrew is an imprint of Elsevier

William Andrew is an imprint of Elsevier
The Boulevard, Langford Lane, Kidlington, Oxford OX5 1GB, UK
225 Wyman Street, Waltham, MA 02451, USA

First edition 1993

British Library Cataloguing in Publication Data
A catalogue record for this book is available from the British Library

Library of Congress Cataloging-in-Publication Data
A catalog record for this book is availabe from the Library of Congress

ISBN: 978-0-12-420038-8

For information on all William Andrew publications
visit our web site at books.elsevier.com

Typeset by TNQ Books and Journals
www.tnq.co.in

Printed and bound in United States of America

14 15 16 17 18   10 9 8 7 6 5 4 3 2 1

This book is dedicated with respect and love to my wife, Giselle T. Reischer CPA, MBA. She encouraged my undertaking its writing, and she may be the only woman financial executive familiar with such words as thixotropic and pozzolanic.

CONTENTS

Dr Richard W. Goodwin, PE is an Environmental Consulting Engineer with over 25 years experience in the waste treatment, disposal, and by-product utilization field including privatization activities in Latin America, Pacific Rim, and Europe. Dr Goodwin's recent activities include assisting resource recovery and independent power owners/operators during permit acquisition and start-up/shake-down phases. He has assisted in achieving by-product utilization for pollution control wastes. His consulting activities with independent power producers achieved fast-track permitting, reduction of operating liabilities, and millions of dollars in cost savings. He is an expert in resource recovery, air pollution control, and residue management. His consulting assignments have included participation in municipal waste combustor (MWC) residue sampling/analysis, disposal, and utilization from operating resource recovery systems throughout the USA. He has assisted the Army Corps of Engineers in the development of a TNT (Trinitrotoluene) waste recovery plant and served as a sole-source consultant to the U.S. Department of Energy regarding the disposal and utilization of residues from advanced combustion processes clean coal technology. He is working with electric utilities on permitting of new coal-fired power plants implementing clean coal technology (consulted to USDOE in this area). Dr Goodwin also consults to investment community including investment bankers, private equity, and venture capital firms. His current work includes anaerobic digestion and energy recovery of biomass; conversion coal-fired plants to biomass and waste-to-energy resource recovery facilities. He consults to electric utilities regarding retrofit and beneficial use of coal combustion residue as a consequence of recent regulatory mandates. He is also a contributor to American Coal Ash Association and Electric Utility Solid Waste Advisory Group regarding EPA's proposed regulation of coal combustion products. He has consulted to electric utilities regarding SDI (Spray Dryer Injection) vs flue gas desulfurization retrofits and natural gas conversions. Consulting to investment firms Shale Gas—Energy Economics. He has been an invited speaker at several state agencies (e.g. NJ, NY, MA) regarding MWC ash management issues. He has represented concerns of National Solid Waste Management Association's Institute of Resource

Recovery to media and federal/state elected, appointed officials. As an invited participant to MacArthur Foundation SIPI TV News, he advocated combustion, cogeneration in resolving energy and waste issues. He has participated in the Congressional and federal review process 1990 Clean Air Act Amendments, and in reauthorization of the Resource Conservation Recovery Act and proposed USEPA Classification of Coal Combustion Residues.

Dr Goodwin has held executive management positions with Research-Cottrell, Chemico Air Pollution Control Corp., and General Electric Environmental Services Inc. He holds degrees from Carnegie Mellon (Tech)/LIU, NYU, and Technion/WOU and Professional Engineer licenses in NY, NJ, and VT. He has lectured at Columbia University, N.Y. Polytechnic Institute, and Rutgers University and is active in several professional societies. Dr Goodwin has presented/published over 70 papers and three books dealing with environmental energy issues (e.g. High-Volume Residues from Combustion Systems).

MANAGEMENT CONSIDERATIONS: COAL AND INCINERATOR COMBUSTOR RESIDUE

The first edition (1) focused on municipal waste combustion residue (MWCR) or incinerator ash; this second edition focuses on coal combustion residues (CCR). There are parallels and differences in discussing these two types of residues.

Similarities of CCR and MWCR

Both residues are released from air pollution control systems—electrostatic precipitators, baghouses, flue gas desulfurization (FGD). Their chemical compositions are similar to Portland cement. They both reflect inherent concrete-like properties achieving liner-like permeability in landfill applications. These residues have been used for beneficial reuse, e.g. FGD by-product gypsum, concrete additives, and construction materials.

Due to the nature of the combusted fuel (municipal solid waste, coal) these residues contain heavy metals—posing environmental concerns.

Environment Concerns of CCR and MWCR—Differences

Twenty years ago when the first edition was published (1993), MWCR's contained heavy metals (lead and cadmium) exceeding the regulatory test level. In a Supreme Court case, Environmental Defense Fund vs City of Chicago, MWCR was deemed a hazardous waste subject to Section C of Resource Conservation and Recovery Act (RCRA) (2). This classification imposes double liners and monitoring wells for all types of landfill or impoundments.

The 2009 proposal by the US Environmental Protection Agency (USEPA) to classify CCR as a hazardous waste was not based upon regulatory testing protocol (as cited in RCRA) but was a knee–jerk reaction to the unfortunate Tennessee Valley (TVA) Kingston ash spill. Recently, USEPA proposed four options to address discharges of CCRs to surface waters. These options offer ranges of numeric limitations of heavy metals. In addition discharges from new coal-fired power plants would also be subject to numeric limitations plus zero discharges. EPA is considering accepted voluntary conversion from wet to dry disposal (3). Such discharge options, however, fail to resolve EPA's proposed classification of CCR as hazardous waste.

Beneficial Reuse Concerns—Differences

Designation of MWCR as hazardous eliminated, for all practical purposes, their beneficial reuse. Some exceptions include Palm Beach County Solid Waste Authority where MWCR is used as a cover for a lined and monitored MSW (Municipal Solid Waste) landfill.

CCR beneficial use has been going on for almost 100 years, e.g. cinder block. The nonprofit American Concrete Institute (ACI), which publishes technical standards, has concerns about EPA's proposed rule. Of some 400 standards and technical documents, 106 would have to be revisited were fly ash ruled to be a hazardous waste. These standards include ACI 232, which outlines the use and application of fly ash, and ACI 318, the model concrete code.

Regulating CCRs as hazardous, based on the Kingston ash spill, is not justified. The cause of the spill was due to TVA's ignoring their geotechnical consultants concern regarding the structural stability of the pond's dike wall. During 2010 the USEPA investigated about 50 CCR impoundments—making recommendations to respective electric utilities retrofitting where required. Thus, a database of engineering criteria exists to develop mandatory guidelines for future impoundments—preventing another dike wall failure. The hazardous waste designation would greatly inhibit the beneficial use of CCRs, e.g. FGD by-product gypsum and fly ash used as additives to construction material, due to potential risk of litigation liability by end-users.

The Kingston ash spill occurred because engineering judgment was ignored—applying an engineering approach to future facilities would not only avoid another accident but continue to use beneficial CCRs based on demonstrated engineering applications.

In 2008 in the US, 136 million tons of CCRs were produced, according to a survey by the American Coal Ash Association. Almost 16 million tons were used in cement and concrete production. A hazardous waste designation for fly ash would stigmatize its use as an ingredient in concrete. The USEPA also proposed to maintain regulation of CCRs as a nonhazardous solid waste. Under both proposals the USEPA would leave in place the Bevill exemption for beneficial uses of coal ash in which coal combustion residuals are used in concrete, cement, wallboard and other contained applications. EPA reasoned that these uses would not be impacted by the proposal.

In June 2011 Veritas Economic Consulting Report said classifying CCR as hazardous waste could cost 183,000–316,000 lost jobs across the nation—41,000–73,000 in the region that includes West Virginia—and $78.9–110 billion over 20 years. These costs reflect both costs to operating coal-fired plants and increased costs of commodities.

Outlook and Perspective—CCR

The outlook for MWCR was set by the court decision; all expanded and new waste-to-energy facilities are subject to Subtitle C regulations—greatly inhibiting siting. The hazardous waste classification imposes a stigma on MWCR and, by implication, waste-to-energy systems. Adoption of a Subtitle C classification would yield similar deleterious effects to coal-fired power plants.

Coal combustion by-products (CCBs) are presently regulated as solid waste (Subtitle D) under the RCRA. Such classification promotes beneficial use by end-users, i.e. mitigating excessive liability. According to the USEPA, about 131 million tons of coal combustion residuals—including 71 million tons of fly ash, 20 million tons of bottom ash and boiler slag, and 40 million tons of FGD material—were generated in the US in 2007. Of this, approximately 36% was disposed of in landfills, 21% was disposed of in surface impoundments, 38% was beneficially reused, and 5% was used as minefill. Stringent regulation, as Subtitle C (Hazardous Waste), would impose a perceived liability upon end-users; greatly reducing beneficial use opportunities. Mandatory use of synthetic liners—would not have prevented dike wall failure and fails to consider inherent engineering characteristics of CCBs (4).

Consider the following uses of power plant wastes to improve how they are land disposed:

Physical Properties of FGD Residue and Fly Ash—Retrofitting Surface Impoundments as Grout to Strengthen Dike Walls

The particle size of FGD residue and fly ash shows this blend could be used as a grout material to stabilize existing CCB surface impoundment dike walls. When used as grout, the blend must be able to penetrate between the interstitial soil spaces. Grouting existing soil dike wall would be about 90% less costly then slurry cutoff wall.

According to AECOM's (Architecture, Engineering, Consulting, Operations and Maintenance) (TVA's forensic geotechnical consultant) June 25, 2009 Summary Report, a combination of the existence of an unusual bottom layer of ash and silt, the high water content of the wet ash, the increasing height of ash, and the construction of the sloping dikes over the wet ash were among the long-evolving conditions that caused a 50-year-old coal ash storage pond breach and subsequent ash spill at TVA's Kingston Fossil Plant on December 22, 2008. Retrofit of surface impoundments, using FGD residue and fly ash and, where required, a cementitious additive would prevent similar dike wall failures.

Residue Management—Placement—Landfill Methodology

The inherent pozzolanic-like behavior of lime-laden CCBs enables achieving improved geotechnical properties i.e. strength, permeability. Achieving liner-like permeabilities, by capitalizing upon CCB's inherent characteristics and applying proper placement control, achieves cost savings of 65% over traditional disposal methods, e.g. synthetic liners (5).

Demonstration Program—Landfill and Surface Impoundment Embankments

Considering the inherent engineering properties of CCBs justifies using this material to form surface impoundment dike walls. Approximately 27.5 million tons of CCBs are retained in surface impoundments. Preventing failure of these dike walls represents a primary issue for discussions between the electric utility industry and regulators. A demonstration program, based on laboratory and bench-scale testing, would indicate industry willingness to address future requirements in a cost-effective manner.

The electric utility industry with their trade and research organizations are urged to commit to conducting such programs (demonstrating the application of CCBs in land disposal); showing a "good faith" effort to cooperate with regulatory and addresses recent CCB disposal upsets.

The USEPA has proposed two classifications for these materials—hazardous or nonhazardous. Beneficial use of ash and FGD sludge (i.e. FGD by-product gypsum) is exempted; the former used as additive to construction materials and the latter used in wallboard and cement production (e.g. Tampa and Seminole Electrics). Since approximately 40% of these wastes are used, if USEPA opts for hazardous waste classification—beneficial use exemption notwithstanding—the threat of potential litigation would defer end-users (e.g. LaFarge Cement, US Gypsum) from reuse.

If the by-product would be regulated as a hazardous material, that would cost industry $1.5 billion a year whereas if it is viewed as a nonhazardous material, it would run $600 million a year. This would result in higher construction material prices and increase electric utilities disposal costs and electricity generation rates.

Dry land-filling CCRs offer a more manageable site option to achieve optimal geotechnical properties, i.e. achieve liner-like permeabilities by capitalizing on the material's inherent pozzolanic properties(5). Ponding of bottom ash requires a more rigorous geotechnical design and monitoring program. Beneficial use of FGD by-product gypsum and fly ash for building materials are ongoing commercial successes.

COMMENTARY

The House of Representatives in October 2011 passed the Coal Residuals Reuse and Management Act (H.R. 2273), a bill that would amend Subtitle D of the Solid Waste Disposal Act (more commonly referred to as the RCRA), by adding Section 4011, Management and Disposal of Coal Combustion Residuals. Sponsored by Sen. Hoeven and others, the Coal Ash Recycling and Oversight Act of 2012 (S.3512) was introduced in the Senate on August 2, 2012. Both bills would essentially create a state-implemented permit program that would manage the disposal of coal combustion residuals. States have permitted cost-effective retrofit use of CCR to strengthen pond dike walls (6). Allowing states to permit CCR disposal and utilization modes, without federal regulations, would facilitate site-specific and cost-effective approaches.

Sound engineering judgment supports the nonhazardous classification of CCRs—based upon decades of operating experience. Consider using CCRs as a retrofit material, i.e. applying a cost-effective, engineering solution.

Richard W. Goodwin
West Palm Beach, FL
August 12, 2013.

REFERENCES

1. Goodwin, R.W.; "Combustion Ash/Residue Management – An Engineering Perspective"; Noyes Publications/William Andrew/Elsevier Publishing; London, England, 1993; (ISBN: 0-8155-1328-3) (Library of Congress Catalog Card No.: 92-47240).
2. Schneider, K.; "Incinerator Operators Say Ruling Will Be Costly"; The New York Times, May 4, 1994.
3. USEPA, News Release; "EPA Proposes to Reduce Toxic Pollutants Discharged into Waterways by Power Plants"; Release Date: 04/19/2013.
4. Goodwin, R.W.; "Avoiding Ash Landfill Operating Mistakes"; Energy Pulse.Net; March 27, 2003.
5. Goodwin, R.W; "Avoiding Costly Mistakes in CCP (Ash, FGF By-Product Gypsum) Landfill Management and By-Product Usage"; CCP Management: Impacts of Regulations and Technological Advances; Electric Utility Consultants Inc.; February 28–29, 2012; Nashville, TN.
6. Goodwin, R.W.; "Management Of Coal Combustion Products (CCP's) – Avoiding Disposal and Utilization Upsets"; EPA Rule Making And Coal Combustion Products; Electric Utility Consultants Inc.; March 14–15, 2011; Denver, CO.

NOTICE

To the best of the Publisher's knowledge the information contained in this book is accurate; however, the Publisher assumes no responsibility nor liability for errors or any consequences arising from the use of the information contained herein. Final determination of the suitability of any information, procedure, or product for use contemplated by any user, and the manner of that use, is the sole responsibility of the user. Due caution should always be exercised in the handling of materials and equipment which could present potential problems. The subject of toxicity relating to combustion ash/residues is controversial, therefore expert advice should be obtained at all times when implementation is being considered for beneficial use of such materials.

All information pertaining to law and regulations is provided for background only. The reader must contact the appropriate legal sources and regulatory authorities for up-to-date regulatory requirements, and their interpretation and implementation.

The book is sold with the understanding that the Publisher is not engaged in rendering legal, engineering, or other professional service. The Publisher is not qualified to judge as to the toxicity of municipal incinerator ash. If advice or other expert assistance is required, the service of a competent professional should be sought.

CHAPTER 1

Fundamental Concepts

Contents

1.1. INTRODUCTION

Generating ash from a beneficial energy combustion process is a thermodynamic certainty; whether the process is burning logs in the fireplace, or burning coal in a power plant, or incinerating municipal refuse in a resource recovery facility. Rather than pose a threat to our environment and act as an impediment to siting waste-to-energy facilities, this ash is not harmful and possesses properties suitable for reuse.

1.2. TYPES OF ASHES

Residues or ash from these waste-to-energy systems may range from 20% to 35% of the municipal solid waste (MSW) feed rate. As depicted by Figure 1.1, these residues are composed of (1) bottom ash (BA), which includes the riddlings or siftings (i.e. the material which falls through the boiler/incinerator) plus grate ash (i.e. material which remains on the grate at the discharge end) and (2) fly ash/scrubber residue (FA/SR), which includes boiler tube or economizer ash (i.e. residue built up on heat transfer surfaces), FA (i.e. finer and lighter particulate matter carried by the flue gas), plus scrubber residue (i.e. reaction products formed by the addition of an alkaline reagent, typically lime, with unreacted lime). FA/SR are collected within

Combustion Ash Residue Management
http://dx.doi.org/10.1016/B978-0-12-420038-8.00001-X

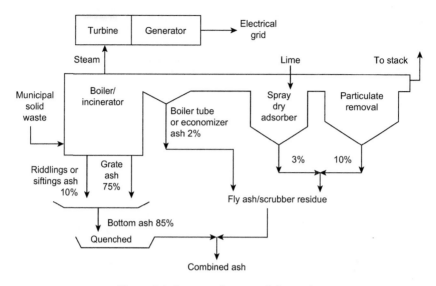

Figure 1.1 *Resource Recovery Schematic.*

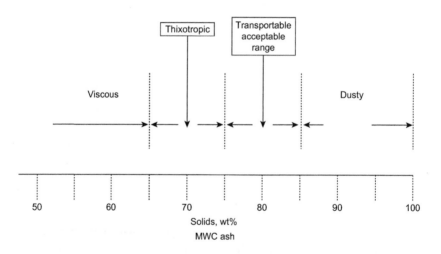

Figure 1.2 *Transportability Schematic.*

the air pollution cleaning (APC) system. A typical APC system consists of an alkaline-based (usually lime) reaction vessel, termed a spray–dryer absorber (SDA), followed by a particulate removal unit, typically a baghouse or electrostatic precipitator (as depicted by Figure 1.1).

As depicted in Figure 1.3, coal combustion products (CCPs) (also called coal combustion residuals) are categorized into four groups, each based on

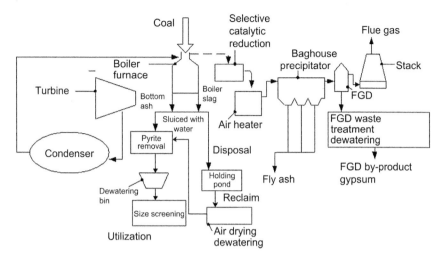

Figure 1.3 *Power Plant Generation of Coal Combustion Products.*

the physical and chemical forms derived from coal combustion methods and emission controls:

FA is captured after coal combustion by filters (baghouses), electrostatic precipitators, and other air pollution control devices. It comprises 60% of all coal combustion waste (labeled here as CCPs), which is most commonly used as a high-performance substitute for Portland cement clinker for Portland cement production. Cements blended with FA are becoming more common. Building material applications range from grouts and masonry products to cellular concrete and roofing tiles. Many asphaltic concrete pavements contain FA. Geotechnical applications include soil stabilization, road base, structural fill, embankments, and mine reclamation. FA also serves as a filler in wood and plastic products, paints, and metal castings.

Flue-gas desulfurization (FGD) materials are produced by chemical "scrubber" emission control systems that remove sulfur and oxides from power plant flue-gas streams. FGD comprises 24% of all coal combustion waste. Residues vary, but the most common are FGD by-product gypsum (or "synthetic" gypsum) and spray-dryer absorbents. FGD gypsum is used in almost 30% of the gypsum panel products manufactured in the United States. It is also used in agricultural applications to treat undesirable soil conditions and to improve crop performance. Other FGD materials are used in mining and land reclamation activities.

Bottom ash and boiler slag can be used as a raw feed for manufacturing Portland cement clinker, as well as for skid control on icy roads. The two materials comprise 12 and 4% of coal combustion waste respectively. These materials are also suitable for geotechnical applications such as structural fills and land reclamation. The physical characteristics of BA and boiler slag lend themselves as replacements for aggregate in flowable fill and in concrete masonry products. Boiler slag is also used for roofing granules and as a blasting grit.

1.3. EFFECT OF AIR POLLUTION CONTROL UPON RESIDUE GENERATION RATES

A typical APC or acid gas cleaning (AGC) system comprises of reacting the incinerator flue gas with lime (usually in an absorption vessel) followed by particulate removal (baghouse or electrostatic precipitator). While these systems are designed to remove acid gases, they increase the waste generation rate due to the production of reaction salts or products, including unreacted lime. The AGC waste is composed of FA and dry flue gas desulfurization (DFGD) (i.e. acid gas treatment) reaction products.

The APC waste (generally referred to as FA) is composed of FA and acid gas treatment reaction products. The BA comprises 75–85% of the total residue, while the FA constitutes the remaining 15–25%. These residues, which may constitute from 20% to 35% of the system feed rate, are in part generated by APC systems. These systems are designed to remove acid gases, but in so doing they increase the waste generation rate.

1.3.1. Utility Waste Analogy—Clean Coal Technology Residue Contribution

Clean coal technologies (CCT's) are discussed since most of them have been applied to MSW waste-to-energy air pollution control systems. Atmospheric fluidized bed combustion (AFBC) involves the combustion, occurring at atmospheric pressure, of fuels in a bed of fluidized sorbent, e.g. limestone; reacting the calcined lime with acid gases (e.g. SO_2, HCl). Calcium spray dryer or DFGD contacts a fine lime or limestone slurry mist and flue gas in a reaction chamber. Acid gases react with this atomized mist to form reaction products and unreacted reagent. Limestone furnace injection is a type of furnace sorbent injection, whereby a pulverized calcium-based sorbent, such as hydrated lime or limestone, is injected directly into the combustion zone above the burners. Calcium duct injection introduces the powdered sorbent (lime or limestone) into the flue-gas ductwork, downstream of the boiler, but upstream of the particulate collector.

Depending on the process and coal sulfur content (plus chlorides for MSW waste-to-energy), the waste generation rate increases by 3.0–4.6 lb residue/lb SO_2 removed (as shown on Table 1.1). A typical 500 MW plant, burning 2.5% sulfur and 29.8% ash coal, without CCT, generates ash at a rate of 102 tons/h (TPH). On average this rate could increase by 41.9 TPH with implementation of CCT.

This 41% increase of residue rate may cause initial concern but understanding the engineering properties mitigates these fears. Unlike wet FGD, CCT residues are dry. They require minimal processing for disposal and utilization. Addition of solubilization water and compaction achieves self-liner, concrete-like field set-up behavior. CCT residues conform to soil stabilization, road base, and structural fill applications without further processing.

Table 1.2 compares the mineralogical composition of utility residues to ash from both mass burn and refuse-derived fuel (RDF) MWC (Municipal Waste Combustor) systems. The considerably higher lime content of MWC residues is due to the higher stoichiometric requirements (i.e. 1.5) of dry acid gas systems; compared to the utility dry FGD stoichiometrics (i.e. approximately 1.2). Typically, recycling of the AGC (FA and reaction products) waste to the dry absorber is not practiced for MWC acid gas treatment systems.

The ash generated from mass burn MWC systems is composed primarily of BA (75–85 wt%) and FA (15–25 wt%). The ash generated from RDF

Table 1.1 CCT Residue Generation Contributions

Applicable (Sulfur Content) and (lb Sorbent Generated per lb SO_2 Removed)	Calcium			Duct Injection
	AFBC	SDA	LFI	
High (4.0%) (lb/lb)	3.0	3.3	4.6	3.9
Medium (2.5%) (lb/lb)	3.0	3.2	4.1	3.5
Low (0.5%) (lb/lb)	3.0	3.0	4.4	3.9

Table 1.2 Comparison of Utility FGD Waste to Residue/APC Waste

Component	Utility DFGD Waste, wt%	Residue/AGC Waste, Mass Burn	RDF, wt%
SiO_2	28	24	37
Al_2O_3	14	6	4
FeO_3	5	3	5
CaO	24	37	43
$CaCl_2$	4	12	7
$CaSO_4/SO_3$	25	18	4

reflects a higher FA (40 wt%) to BA (60 wt%) distribution; due to RDF's suspension firing. RDF ash reflects a higher lime content than mass burn residues.

1.4. IN-PLANT ASH HANDLING

Depending upon the design of the ash handling system the FA/AGC waste may be combined with the BA. Since the BA is quenched, the resultant blend will contain water. This mixture of ash and AGC wastes is conveyed to a disposal site. Since this material may be thixotropic and contains considerable lime, the solid content should be maintained to ensure optimal transportability characteristics, i.e. (1) prevention of fugitive dusting; (2) elimination of thixotropic behavior; and (3) avoidance of premature pozzolanic reaction. Typically, the transportable solid content could range from 75 to 85 wt% to satisfy these transportability criteria.

Coal BA and boiler slag are the coarse, granular, incombustible by-products that are collected from the bottom of furnaces that burn coal for the generation of steam.

The most common type of coal-burning furnace in the electric utility industry is the dry, bottom pulverized coal boiler. When pulverized coal is burned in a dry, bottom boiler, about 80% of the unburned material or ash is entrained in the flue gas and is captured and recovered as FA. The remaining 20% of the ash is dry BA, a dark gray, granular, porous, predominantly sand size minus 12.7 mm (0.5 in) material that is collected in a water-filled hopper at the bottom of the furnace.

There are two types of wet-bottom boilers: the slag-tap boiler and the cyclone boiler. The slag-tap boiler burns pulverized coal and the cyclone boiler burns crushed coal. In each type, the BA is kept in a molten state and tapped off as a liquid. Both boiler types have a solid base with an orifice that can be opened to permit the molten ash that has collected at the base to flow into the ash hopper below. The ash hopper in wet-bottom furnaces contains quenching water. When the molten slag comes in contact with the quenching water, it fractures instantly, crystallizes, and forms pellets. The resulting boiler slag, often referred to as "black beauty", is a coarse, hard, black, angular, and glassy material.

When pulverized coal is burned in a slag-tap furnace, as much as 50% of the ash is retained in the furnace as boiler slag. In a cyclone furnace, which burns crushed coal, some 70–80% of the ash is retained as boiler slag, with only 20–30% leaving the furnace in the form of FA.

Wet-bottom boiler slag is a term that describes the molten condition of the ash as it is drawn from the bottom of the slag-tap or cyclone furnaces. At intervals, high-pressure water jets wash the boiler slag from the hopper pit into a sluiceway which then conveys it to a collection basin for dewatering, possible crushing or screening, and either disposal or reuse.

More than 80% of BA systems remain wet—posing an environmental risk as evidence by TVA Kingston Ash Spill. The US Environmental Protection Agency (EPA) may require retrofitting these ponds to dry disposal. In 2010 the EPA estimated that the cost for such conversion could exceed $20 billion industry-wide. About 40% of coal ash is recycled. More than 300 coal-fired plants in United States, landfill this ash. Nearly 150 plants use off-site commercial landfills as their disposal means. Nearly 160 US plants use coal ponds for disposal (1).

FGD waste is pumped from the scrubbers or FGD systems and dewatering via cyclones and vacuum filters. Sometimes the FA and FGD waste are blended to facilitate landfill. This option is less desirable than the beneficial use of FA and FGD by-product gypsum.

1.4.1. Fugitive Dusting

Considerable environmental rhetoric has focused on perceived issues of fugitive dusting—during transport to placement. Based on empirical studies, the addition of dust suppression water eliminates airborne dusting when the ash remains as a soil-like material. If the ash is agglomerated to accommodate reuse applications, fugitive dusting only occurs under loss of structural integrity. The engineering approach provides insight into the fugitive dusting issue by offering an empirical approach. The durability tests, discussed within chapter five's proposed protocol, depict worst-case concerns and provide simulatory testing assurance.

Figure 1.2 schematically depicts the phases of transportability upset conditions—identifying a finite range of solid contents to avoid these undesirable modes. The MWC facility's water balance can be optimized to ensure attaining the desirable solids content, i.e. transportability range. To avoid excessive dusting within the facility itself, good engineering practice suggests implementing such measures as covering conveyors and use of moisture-conditioners or dustless unloaders (commonly used at coal-fired power plants). During transport closed or covered vehicles should be employed. In-plant fugitive dust control reflects good plant housekeeping practice. These recommended measures typify those used at modern combustion energy facilities.

REFERENCES

1. Russell, R.; "Coal Ash Handling & Storage: Shifting Direction"; Power Engineering; Vol. 117, No. 2, February 2013; pp 22–29.

FURTHER READING

Goodwin, R.W.; Schuetzenduebel, W.G.; "Residues from Mass Burn Systems: Testing, Disposal and Utilization Issues"; Proceedings NYS Legislative Commission's Solid Waste Mgt and Materials Policy Conference; February 1987; NYC.

Goodwin, R.W.; "Concrete-Like Non-Hazardous Behavior MSW Ash with Acid Gas Reaction Products"; Proceedings of the 1988 Conference on Solid Waste Management and Materials Policy; NY State Legislative Commission Solid Waste Mgt.; January 27–30, 1988; Penta Hotel; NYC, NY.

Goodwin, R.W.; "Coal and Incinerator Ash in Pozzolanic Reaction Applications"; presented at MSW Ash Utilization Conference sponsored by Resource Recovery Report; October 13–14, 1988; Penn Tower Hotel; Philadelphia, PA.

Goodwin, R.W.; "Design Optimization of a Flue Gas Desulfurization Sludge Handling System"; Proceedings of Columbia University Seminar on Pollution and Water Resources; Vol. IX, 975, 1978; N.J. Dept. of Environmental Protection, Bureau of Geology & Topography; Bulletin 75-C; pp I1–I18.

Korn, J.L.; Huitric, R.L.; "Combined Ash Treatment for the Commerce Refuse-to-Energy Facility"; presented at ASTSWMO National Solid Waste Forum, July 20–22, 1992; Portland, OR.

Schuetzenduebel, W.G.; Nobles, W.C.; "Hennepin County Resource Recovery Facility"; American Society of Civil Engineers Journal of Energy Engineering; Vol. 117, No. 1, April 1991; pp 1–17.

CHAPTER 2

Governmental Regulations

Contents

2.1. INTRODUCTION

As an aid to developing countries implementing waste-to-energy systems (WTE), some parts of this book's first edition has been retained.

As a background, the method of sampling and types of tests applied to municipal waste combustor (MWC) residues are included (1). These issues are addressed in terms of actual practice. The state regulatory agencies issue environmental permits for the construction/operation of the resource recovery facilities. As a part of this permit, the MWC residue had been typically tested in accordance with the United States Environmental Protection Agency's (USEPA's) Extraction Procedure Toxicity (EP Tox) Method (2,3), which has been replaced by the Toxicity Characteristic Leachate Procedure (TCLP) in September 1990. Due to the wide variation of the MSW feedstock, the ash should be composite sampled. Even under such a scenario, studies have shown that MWC ash exceeds regulatory requirements as prescribed by the provisions of the Resource Conservation and Recovery Act (RCRA) of 1976 and its subsequent revisions.

2.2. PRIOR PROPOSED LEGISLATION

Previously three proposed pieces of legislation could have impacted the present regulation of MWC ash: (1) Senate Bill (S. 1894, S. 2458); (2) Florio Bill (H.R. 4902); and (3) Luken Bill (H.R. 4357). Rather than discussing these proposals in detail, Kochieisen (4) has provided a comprehensive comparative analysis. Common points of this analysis reflect regulatory approaches, which justify commentary.

These common points are listed, as follows:
* minimum requirements of MWC residue disposal monofills; and
* design requirements.

Such examples of proposed legislative initiatives failed to consider the inherent properties of MWC residue and the application of civil engineering and concrete chemistry principles. Such legislation relied on just the lab leachate data, which subsequently, has grossly overstated what actually happens in the field. Reliance on misleading lab results and premature excessive regulation by some states have proven to be technically unjustified, exerting unwarranted economic penalties on WTE facilities.

As noted previously, the Environmental Defense Fund (EDF) vs City of Chicago changed the classification of MWC residues from a nonhazardous classification to hazardous waste. As such MWC residues are subject to regulatory testing to determine if the material from a specific WTE exceeds RCRA criteria for a hazardous material. Most WTE units were installed prior to the court decision so each state regulated the disposal of MWC residues. Consider the following WTE installations and how respective states permitted MWC residue disposal. Lee County FL, Palm Beach County, FL, Warren County NY, Montgomery County MD.

Since the MWC residues are periodically monitored, they are disposed of in a landfill based on hazardous waste design even if the test result did not exceed hazardous material limits. The operators and state permitting agency, assumed a worst-case scenario requiring more stringent landfill design provisions. Typically MWC residues are disposed of in a monofill i.e. fly ash only. The monofill is equipped with liners and monitoring wells.

The USEPA developed a test called the Toxicity Characteristic Leachate Procedure (TCLP) that subjects ash to acidic liquid, causing metals to leach from the material. If metals leach in amounts greater than a fraction of a percent, the ash is considered hazardous. Years of testing ash from every WTE facility in the country, has proven it to be safe for disposal and reuse (5).

2.2.1. Proposed Legislation of Coal Combustion Residues

Coal combustion residuals, often referred to as CCRs, are the residuals produced from burning coal for the generation of electricity. CCRs represent one of the largest waste streams in the United States. According to the EPA, CCRs contain low concentrations of metals such as arsenic, cadmium, lead, and mercury. The Bevill amendment of the Solid Waste Disposal Act Amendments of 1980 included CCRs as a "special waste" considered temporarily exempt from regulation as hazardous waste under subtitle C of the RCRA.

In December 2008, an estimated 5.4 million cubic yards of coal ash sludge were accidentally released from a disposal containment dike at a Kingston, Tennessee power plant. This incident prompted to evaluate the need for further regulation—this time in the form of the June 21, 2010 proposed rule to regulate CCR disposal. In the proposed rule, EPA proposed two options for the management of CCRs. Under the first proposal, EPA would list these residuals as special wastes subject to regulation under subtitle C of RCRA, when destined for disposal in landfills or surface impoundments. Under the second proposal, EPA would regulate coal ash under subtitle D of RCRA, the section for nonhazardous wastes. The proposal would not alter the regulatory status of coal ash that is beneficially used. However, EPA has identified concerns with some beneficial uses of CCRs in unencapsulated form, such as the use of CCRs in road embankments and agricultural applications.

Under both approaches, the agency would leave in place the Bevill exemption for beneficial uses of coal ash in which CCRs are used in concrete, cement, wallboard, and other contained applications. These uses would not be impacted by the proposal.

The public comment period on the proposed rule closed on November 19, 2010. The proposal drew more than 450,000 comments, including the strong opposition from industry who charge that the option for regulation under strict subtitle C requirements of RCRA (usually reserved for hazardous waste) would impose a stigma on CCRs and lead to a reduction in the beneficial reuse of the material in cement, gypsum, and other products. Environmental groups argue that regulation as a solid waste under subtitle D of RCRA would not adequately limit releases, does not allow for federal enforcement of disposal requirements, or require states to modify their regulations.

In October 2011, EPA released a related Notice of Date Availability (NODA). This NODA announced and invited comment, through November 14, 2011, on additional information obtained by EPA in conjunction with the June 21, 2010 proposed rule. This additional information was categorized as follows: chemical constituent data from CCRs; facility and waste management unit data; information on additional alleged damage cases; adequacy of state programs; and beneficial use.

Exadministrator Lisa Jackson has said that the agency plans to finalize the rule in late 2012 after completing a risk analysis of coal ash reuse in products. The risk analysis is being conducted in response to a March 2011 EPA Inspector General Report, which concluded that EPA promoted the beneficial use of coal ash products and application, such as in gypsum panels and concrete, with incomplete risk information. As of this writing, EPA has given no clear indication of their final decision. Lacking such direction, The States regulations provide guidance to the disposal and possible beneficial use of CCRs.

The draft regulations target a five-to-seven year implementation window to phase out the unlined ponds. The EPA estimate for converting to lined systems is between $587 million and $1.5 billion industry-wide. A June 2011 Veritas Economic Consulting Report said classifying coal ash as hazardous waste could cost 183,000–316,000 lost jobs across the nation—41,000 to 73,000 in the region that includes West Virginia—and $78.9–$110 billion over 20 years.

Besides these economic consequences, other implications include:
- Two to four years for siting, permitting, and construction of the initial cell of a new landfill, plus leachate and runoff/run-on controls and access roads.
- Two years for adding a new cell to an existing landfill requiring upgrades to the disposal facility.
- 6–12 months to permit and physically close an existing 50-acre surface impoundment.
- Two to three years to convert from a wet to a dry ash-handling system.

In 2011 the U.S. House of Representatives voted to instruct the Transportation Conference Committee to keep language in the highway bill that would block the EPA from designating coal ash as hazardous waste. The language—in the form of an amendment by West Virginia Republican Representative David McKinley—was included in the House's Surface and Transportation Extension Act of 2012. H.R. 1391, the "Recycling Coal Combustion Residuals Accessibility Act of 2011", would prevent the USEPA from regulating CCRs and often called "coal ash", under the hazardous waste subtitle of the RCRA. A U.S. House of Representatives Subcommittee on June 21 approved a bill that would prevent the USEPA from regulating coal ash disposal as a "hazardous waste", while simultaneously directing states to enact enforceable permit programs.

The House Energy and Commerce Subcommittee on Environment and the Economy approved H.R. 2273—the "Coal Residuals Reuse and Management Act". The bill, sponsored by Rep. David McKinley (R–WV) effectively replaced H.R. 1391, which Rep. McKinley filed in April.

Rep. McKinley's original bill simply blocked EPA from regulating coal ash as a hazardous waste under Subtitle C of the RCRA. The new bill advocated a state-administered permit program to create enforceable requirements for groundwater monitoring, lining of landfills, corrective action when environmental damage occurs, and structural criteria. H.R. 2273 also provided that if a state could not implement the permit programs, the federal EPA would have authority to do so. This bill would allow the states to continue operating their existing coal ash residue programs. The bill would regulate CCR under Subtitle D when it is recycled or disposed of in a manner acceptable to State Regulators.

A bill introduced in the U.S. Senate would set up a state permitting program for CCR and ensure their storage sites have requirements for groundwater monitoring and protective lining.

Introduced by Sen. John Hoeven, R–N.D., Sen. Kent Conrad, D–N.D., and Sen. Max Baucus, D–Mont., the Coal Ash Recycling and Oversight Act of 2012 was criticized by various environmental groups. In a joint press release, the senators said legislation would ensure safe and efficient recycling of coal ash into value construction materials. Under the measure, states could set up their own permitting processes based on various federal regulations and could grant oversight to the USEPA if they didn't want to have their own process.

Land-filling Coal Combustion Residues offer a more manageable site option to achieve optimal geotechnical properties i.e. achieve liner-like permeabilities by capitalizing on the material's inherent pozzolanic properties. Ponding of bottom ash requires a more rigorous geotechnical design and monitoring program. Beneficial use of FGD (Flue Gas Desulfurization) by-product gypsum and fly ash for building materials are on-going commercial successes.

2.2.2. Regulatory Lessons Learned from MWC Residues— Applicable to USEPA Proposed Regulatory Options

The EPA issued a proposed rule on June 21, 2010, in which it outlined two approaches: One would regulate combustion residuals under Subtitle C of the RCRA as a "special waste" and one would regulate these wastes under Subtitle D of RCRA as a nonhazardous waste. The EPA's timeline notes that the final rule was sent to the White House's Office of Management and Budget in March 2012, but it has no projected publication date for the final rule.

Since MWC residues are subject to hazardous waste testing, end-users have declined beneficial use options. In accordance with the federal law, WTE ash is tested to ensure it is nonhazardous (e.g. road construction, concrete block etc). The implication of hazardous waste creates the potential of litigation and risk assignment that private sector firms avoid—especially in the litigious USA society. The lesson of subjecting CCR to regulatory testing would have the same effect so the USEPA, lacking substantive data, should not regulate these residues as hazardous material. The effect would have far-reaching consequences since CCRs have been beneficially used for decades.

2.2.3. Regulatory Laboratory Leachate Tests

A comparison of the content (i.e. mg/kg) to lab leachate tests (Table 2.1) shows a direct relationship between heavy metal specie content and concentration. Similar direct relationships between heavy metal content and lab leachate concentration from the same composite samples are shown on Table 2.2, derived from the Coalition on Resource Recovery and the Environment (CORRE)/USEPA study of five ash monofills (6). This table also shows a noncorrelation of these results to field leachates, casting doubt on the efficacy of establishing regulations (i.e. RCRA Subtitle C Hazardous Waste) based on nonempirically reflective lab tests.

Lacking evidence to the contrary, the public and media have viewed MWC residue as "toxic" (7). Elected officials, similarly, rely on such laboratory data to shape their perspective (8). "With respect to my stand on the disposal of incinerator ash, I believe that incinerator ash that is tested as

Table 2.1 RCRA Heavy Metals: Content, EP Tox, and TCLP Analyses of MWC Combined Ash

RCRA Heavy Metal	mg/kg (a)	EP TOX (mg/l) (b)	TCLP (mg/l) (c)	Maximum Allowable Limits (mg/l)
Arsenic [As]	15–56	ND–0.031	ND–0.060	5.0
Barium [Ba]	193–1000	0.023–0.455	0.110–1.850	100.0
Cadmium [Cd]	1.8–152	0.020–1.200	ND–1.560	1.0
Chromium [Cr]	22–1070	ND–0.086	ND–0.799	5.0
Lead [Pb]	561–22400	ND–19.700	ND–26.400	5.0
Mercury [Hg]	0.55–25.1	ND–0.023	ND–0.001	0.2
Selenium [Se]	ND–50	ND–0.002	ND–0.007	1.0
Silver [Ag]	4.1–13.0	NA–ND	NA–ND	5.0

NA = Not Analyzed; ND = Not Detected; REFS: (a) 6, 10, 24; (b) 26, 25; (c) 26, 6, 15

Table 2.2 CORRE/USEPA RCRA Heavy Metals Content vs Lab/Field Leachate

RCRA Heavy Metal	mg/kg	EP TOX (mg/l)	TCLP (mg/l)	Field Leachate	PDWS (mg/l)
Arsenic [As]	15–56	ND–0.031	ND	ND–0.400	0.05
Barium [Ba]	260–1000	0.173–0.455	0.161–1.850	ND–9.220	1.00
Cadmium [Ca]	18–152	0.025–1.000	ND–0.833	ND–0.004	0.01
Chromium [Cr]	53–665	ND–0.086	ND–0.289	ND–0.032	0.05
Lead [Pb]	1070–1740	ND–15400	ND–8.84	ND–0.054	0.05
Mercury [Hg]	3.2–25.1	ND–0.023	ND–0.005	ND	0.002
Selenium [Se]	ND–5.7	ND	ND	ND–0.340★	0.01
Silver [Ag]	4.1–13	ND	ND	ND	0.05

Notes: PDWS = Primary Drinking Water Standards; NA = Not Analyzed; ND = Not Detected; ★Not found in bulk analyses, i.e., mg/kg; CORRE = Coalition on Resource Recovery and the Environment

hazardous must be treated like any other hazardous material" (9). "…public concern over…toxic ash can not be overlooked…many questions still remain concerning public health risks…(of) disposal of incinerator ash, which contains everything that is not combustible, most notably heavy metals…" (10).

The USEPA has applied two laboratory leachate tests, EP Tox and TCLP, for hazardous classification. Several investigators have analyzed MWC combined ash (i.e. bottom ash, fly ash, and reaction products, if lime is added to flue gas) for these RCRA heavy metals, as summarized by Tables 2.1 and 2.2. Considering such laboratory results, exceeding allowable limitations for cadmium and lead, would classify the ash as hazardous.

Reliance on just laboratory leachate data has led to state regulations imposing economic burdens on resource recovery projects. Costly hazardous waste multilining requirements have been implemented in the northeast ranging from $250,000 to $500,000 per acre (11). Translation of such costs into higher municipal taxes is beyond this book's scope but their burden should be realized. Analyzing available field data suggests that sole reliance on lab data and subsequent regulatory mandates could be considered excessive. Failure to appreciate field data further exacerbates the public mind-set opposing MWC projects, specifically, and heavy industrial projects, in general. Applying such a hazardous waste stigma may be inappropriate when the chemical behavior of the MWC ashes is used to explain relatively environmentally benign field behavior. Thus the regulatory proposals reflect just the issues concerned with the ash testing and design of ash landfills. These proposals fail to discuss the management of the MWC ash.

2.3. CURRENT FEDERAL REGULATION OF MUNICIPAL WASTE COMBUSTION RESIDUES

On May 2, 1994, in the case of City of Chicago vs EDF, the U.S. Supreme Court ruled that MWC residues that exhibit a hazardous waste characteristic are not exempt from regulation as a Solid Waste Programs—11 hazardous wastes under RCRA. Owners and operators of WTE facilities are now required to determine whether their ash is hazardous. The MWC residue must be tested in accordance with USEPA requirements most importantly TCLP, i.e. heavy metals. Facilities generating MWC residue, which is a hazardous waste, must be managed in accordance with RCRA hazardous waste regulations. If, upon testing, the MWC residue is not a hazardous waste, it may be disposed of in a MSW Class I landfill that meets applicable RCRA standards. Because of the substantial confusion caused by the evolution of the federal regulatory interpretation, EPA has developed a strategy to gradually phase in the hazardous waste regulations and enforcement provisions that now apply to the WTE facilities. In the City of Chicago case, the Supreme Court issued a narrowly focused opinion that RCRA §3001(i) does not exempt MWC residues. The Court did not address the issue of the precise point at which regulation of waste must begin, and §3001(i) does not expressly address the issue. In an effort to provide some guidance to the regulated community, EPA published a Notice of Statutory interpretation entitled "Determination of Point at Which Subtitle C Jurisdiction Begins for Municipal Combustion Ash at WTE Facilities" on February 3, 1995.

EPA interprets RCRA §3001(i) to first impose hazardous waste regulation at the point that the ash leaves the "resource recovery facility," defined as the combustion building, including connected air pollution control equipment. Consequently, the point at which the hazardous waste determination for the MWC residue should be made and when Land Disposal Restrictions standards, once promulgated, will begin to apply is the point at which the MWC residue exits the combustion building following the combustion and air pollution control processes. This interpretation is critical, because it means that many facilities will be able to test their MWC residue after the fly ash and the bottom ash have been combined. Often when fly ash that exhibits the toxicity characteristic is combined with bottom ash, the resulting mixture no longer exhibits a characteristic of hazardous waste, and would not be regulated as such. In the February notice, EPA asserted that if it comes to EPA's attention that MWC residue is being managed or disposed of in a manner that is not protective of human health and the environment, the Agency may consider issuing management guidelines or promulgating additional regulations to address those situations. In addition, at individual sites, if the disposal of MWC residue presents an imminent and substantial endangerment to human health and the environment, EPA may invoke RCRA §7003 authorities to require responsible parties to undertake appropriate action (12).

2.4. STATE REGULATION OF MUNICIPAL WASTE COMBUSTION RESIDUES

If regulatory tests show that the residue exceeds RCRA standards, it may be landfilled but only in accordance with a hazardous landfill requirement, e.g. double-liners, monitoring wells etc. If tests show compliance with RCRA standards a nonhazardous monofill design would be acceptable. In Warren County NJ, the residue was periodically tested and deemed nonhazardous— disposed at a nonhazardous landfill. In Palm Beach County FL, the residue is used as a daily cover for MSW landfill i.e. nonhazardous landfill.

According to Association State and Territorial Solid Waste Management Officials' (ASTSWMO's) 2006 Beneficial Use Survey Report (http:// www.astswmo.org/files/publications/solidwaste/2007BUSurveyReport11-3 0-07.pdf) 10 states have approved beneficial use requests for the reuse of WTE ash. (This may also include non-MSW ash.) They are Florida, Hawaii, Maryland, Massachusetts, Mississippi, New Hampshire, New Jersey, New York, Pennsylvania, and Tennessee (Tennessee's approval was later revoked).

Florida, Massachusetts, and Pennsylvania have approved its use in the manufacture of asphalt; New Hampshire has approved it as an aggregate in asphalt paving. Pennsylvania approved its use as a construction material and in road construction. Florida, Massachusetts, Mississippi, and New York have approved it for various landfill uses.

According to the survey, the use of this ash has not been approved, or is no longer eligible for approval, (1) for asphalt production (Pennsylvania); (2) for fill (Florida, New Hampshire, and Tennessee); (3) as an additive in a concrete road (Michigan); and (4) in construction (Mississippi).

2.4.1. Lee County, FL

Nonferrous materials separated from the residue stream are sold to the secondary metal market. Lee County manages its own ash disposal facility at the Lee/Hendry Landfill owned by Lee County. The County developed ash monofills for disposal of residue from the WTE operation. In addition, the County uses ash residue as a daily cover material for portions of the landfill reserved for municipal solid waste disposal, but the WTE expansion project has greatly reduced this requirement. Ash residue is used for cover in C&D disposal areas. Prior to leaving the WTE complex, any ash remaining after the combustion process is "stabilized" with dolomite lime to buffer the residue to an ideal pH range. The dolomitic lime system was converted from a manual feedbag operation to an automatic system using pneumatic transport as part of the WTE expansion project. The combined fly ash and bottom ash is further processed for ferrous and nonferrous recovery prior to loadout and transport to the landfill (Solid Waste Management in Florida DEP 1999).

2.4.2. Palm Beach County, FL

Bottom ash is collected from beneath the combustion chamber. Ferrous and nonferrous metals are recovered from the bottom ash. Fly ash and bottom ash or MWC residues are combined within the plant and delivered to the Class I (nonhazardous) landfill for disposal.

According to (13), use of the MWC residue as an intermediate cover in the Class I Landfill will not present a significant threat to human health from direct exposure. At least two feet of clean fill is placed on top of the residue as specified in Section 4.2 of the Florida Department of Environmental Protection MWC residue reuse guidance document. The Class I Landfill is a secure area, with controlled access to the general public. Perimeter fencing and security is in place to preclude unauthorized access. Landfill operators are trained in proper ash-handling techniques and are protected by the rules of the Occupational Safety and Health Act. These in-place institutional controls are sufficient to ensure that the ash material will not be disturbed in the future.

Palm Beach County SWA (Solid Waste Authority) states that use of the residue, when limited to within the lined area of the landfill, groundwater contamination concerns are minimal. The Class I Landfill is equipped with a leachate collection and recovery system. Designed to collect and properly dispose of it.

2.4.3. Warren County, NJ

Bottom ash as fly ash are combined and conveyed to the ash dischargers, which contain water to quench the hot ash. Cooled ash settles to the bottom of the discharger and is pushed by a hydraulic ram onto the main ash conveyor. The resulting combined ash or MWC residue stream is conveyed to the ash storage building and then loaded into dump trucks and taken to the county landfill where it is used as beneficial reuse/daily cover. The facility utilizes a ferrous metal recovery system to recycle thousands of tons of metal each year.

2.4.4. Montgomery County, MD

The Out-of-County Waste Transportation and Disposal Program includes a 15-year contract (with a 5-year extension option) with Brunswick Waste Management Facility, Inc. for waste transportation and disposal services. The contract was executed in June 1997 and the initial 15-year contract term runs through June 30, 2012. Services began October 20, 1997.

The original objectives of the contract and associated administration and support requirements were to assure that MWC residue from the County's WTE facility generated in Montgomery County and managed at County solid waste facilities was transported to the contractor's facility in Brunswick County, Virginia, and disposed in the County's contracted dedicated disposal cell.

2.4.5. Regulatory Recognition of MWC Inherent Pozzolanic Behavior

Recent field studies of MWC ash landfills strongly support the relatively benign characteristics of this ash. Reference (14) states that "the leachates… are close to being acceptable for drinking water use, as far as the metals are concerned". Both the public and regulatory community have focused on the results of laboratory tests (e.g. EP Tox, TCLP) to predict MSW ashes' leaching. But these tests do not reflect field leachates results; "leachate from the disposal sites tested out below the level that those two tests deem hazardous" (15).

Although EP Tox results have shown excessive levels of Pb and Cd, the presence of unreacted lime (from the APC system) could account for significant reductions of constituents. The author has contended that such ash should be deemed pozzolanic and recognized by regulatory authorities (16). Recently some state regulatory agencies have recognized this behavior and incorporated it within their classification of MWC ash landfills. The California Department of Health Services concluded that MWC "…ash possesses intrinsic physical and chemical properties rendering it insignificant as a hazard to human health and safety, livestock, and wildlife". The "intrinsic property" is the formation of a "lime/pozzolan mixture" so that when "compacted (the) ash forms a hard, nonerodible surface" (17). The necessity to further treat MWC ash containing excess lime from scrubbers "…contain sufficient pozzolanic properties which…will result in some hardening of the ashes without additional additives" (18).

2.5. STATE REGULATION OF COAL COMBUSTION RESIDUES

As this second edition is written, the USEPA has not finalized its classification of CCR. So the prevailing classification of CCRs (under the Resource and Recovery Act) is nonhazardous, Class D material. This section offers insight into how most affected states regulate CCR.

Definitions and management of CCRs vary widely among states, but mostly take an active role in regulating the material, a group of solid waste regulators said in a recent survey.

ASTSWMO, in 2006 Survey, reported that of the 46 states that responded to the survey, 48% have a statutory or regulatory definition of ash from coal-fired power plants. Seventy-six percent of states said that they restrict the beneficial reuse of coal ash through a variety of methods.

2.5.1. North Carolina

North Carolina, along with EPA and other states across the country, promotes the reuse and recycling of solid waste. A significant percentage of CCRs generated in North Carolina are reused or recycled. Regulation of CCRs destined for disposal as a hazardous waste under RCRA will have a chilling effect on an important industry that is reducing the volume of CCRs disposed. There are options available to address risks from the disposal of CCRs that avoid this undesired effect.

North Carolina Department Waste Management proposes that EPA regulate CCRs under Subtitle D of RCRA with modifications to the current proposal. Rule changes—and if necessary statutory changes— should be made to make federally approved state permitting programs the foundation for regulating CCR disposal. EPA should also provide financial incentives for states to implement federal criteria through state solid waste programs.

(North Carolina Department of Environment and Natural Resources, Division of Waste Management; November 17, 2010.)

2.5.2. Oklahoma

Oklahoma regulations adopt by reference the federal regulation that exempts CCRs (including fly ash, bottom ash, slag, and flue gas emission control waste generated primarily from the combustion of coal) from classification as hazardous waste. Reuse of coal combustion by-products (CCBs) is not specifically authorized under Oklahoma law. CCRs are exempt from solid waste regulations if constructively reutilized in approved mine applications. Note: Fly ash and bottom ash generated outside the state must be constructively reutilized or disposed only in active or inactive mining operations subject to state laws and regulations.

Oklahoma regulations adopt by reference the federal regulation, which exempts CCRs from classification as hazardous waste. Exempt from hazardous waste regulation are fly ash, bottom ash, slag, and flue gas emission control waste generated primarily from the combustion of coal. OAK. REG. 252:205-3-2(c); 40 CFR 261.4.

According to a February 29, 2000, CCR Policy Statement, the Oklahoma Department of Environmental Quality currently allows the following uses of CCPs without prior approval when used according to the applicable standard (generally American Society Testing Materials): cementitious material production, daily landfill cover (with permit modification), as manufactured product, road base, road surfacing material, solidification/chemical fixation, deicing, soil stabilization, subgrade treatment, engineering applications, and mine reclamation. Records of reuse and disposal of CCRs must be kept and storage shall be in an environmentally appropriate manner to prevent releases to the environment.

Note: Fly ash and bottom ash over 200 tons and generated outside the state must be constructively reutilized or disposed only in active or inactive mining operations subject to state laws and regulations (OKLA. STAT. 27A §2-10-801(F)). CCRs are exempt from solid waste regulations if constructively reutilized in approved mine applications (OKLA. STAT. 45-12-§940).

2.5.3. Florida

Under Florida regulations, fly ash, bottom ash, slag, and flue gas emission control waste generated primarily from the combustion of coal or other fossil fuels are exempt from regulation as hazardous waste. Reuse of CCRs is not specifically authorized under Florida law.

Florida regulations adopt the current federal regulations which exempt fly ash, bottom ash, slag, and flue gas emission control (FGD) waste generated primarily from the combustion of coal or other fossil fuels from regulation as hazardous waste (FAC 62-730.030). CCRs are regulated as solid waste if disposed of (FAC 62-701) and may be regulated as industrial by-products if the CCRs are utilized within one year, if there is no release or threat of release into the environment, and if the facility is registered with the Department of Environmental Protection to allow for such recovery of CCRs.

Ash residue from CCRs for use in concrete is specifically authorized under Florida law (FLA. STAT. 336.044).

2.5.4. Ohio

Under Ohio regulations, fly ash, bottom ash, slag, and flue gas emission control waste generated primarily from the combustion of coal or other fossil fuels are exempt from regulation as hazardous waste. Reuse of CCRs is not specifically authorized under Ohio law; however, reuse of "nontoxic" CCRs is authorized under policy documents issued by the Ohio Environmental Protection Agency (OEPA), Division of Water Quality. Nontoxic CCRs may be reused (1) as a raw material in manufacturing a final product; (2) as a stabilization/solidification agent for other wastes that will be disposed; (3) as a part of a composting process; (4) in uses subject to USEPA procurement guidelines; (5) for extraction or recovery of materials and compounds in CCRs; (6) as an antiskid material or road preparation material; (7) for use in mine subsidence stabilization, mine fire control, and mine sealing; (8) as an additive in commercial soil blending operations, where the product will be used for growth of ornamentals (no food crops or grazed land); (9) as daily cover at a landfill; (10) as structural fill, defined as an engineered use of waste material as a building or equipment supportive base or foundation and does not include valley fills or filling of open pits from coal or industrial mineral mining; (11) as pipe bedding, for uses other than transport of potable water; (12) as a construction material for roads or parking lots (subbase or final cover); and (13) other single beneficial uses of less than 200 tons. Certain guidelines may apply to the above uses.

(Ohio Environmental Protection Agency (OEPA); Policy No. DSW400.007; November 7, 1994.)

2.5.5. Pennsylvania

Under Pennsylvania regulations, fly ash, bottom ash, slag, and flue gas emission control waste generated primarily from the combustion of coal or other fossil fuels are exempt from regulation as hazardous waste. Coal ash (defined as fly ash, bottom ash, and boiler slag resulting from the combustion of coal) is regulated under the Solid Waste Management Act and the residual waste management regulations, which authorize the beneficial use of coal ash (1) as a structural fill; (2) as a soil substitute or additive; (3) for reclamation at an active surface coal mine site, a coal refuse reprocessing site, or a coal refuse disposal site; (4) for reclamation at an abandoned coal or an abandoned noncoal (industrial mineral) mine site; (5) in the manufacture of concrete; (6) for the extraction or recovery of one or more materials and compounds contained within the coal ash; (7) as an antiskid material or road surface preparation material (bottom ash or boiler slag only); (8) as a raw material for a product with commercial value; (9) for mine subsidence control, mine fire control, and mine sealing; (10) as a drainage material or pipe bedding; and (11) as a stabilized product where the physical or chemical characteristics are altered prior to use or during placement so that the potential of the coal ash to leach constituents into the environment is reduced. All of these uses must comply with specified State regulations.

Beneficial use of CCR was implemented through Department of Environmental Protection ("DEP") guidelines under the residual waste management regulations, 25 Pa. Code Chapter 287, which were amended in July 1992 to include the beneficial use of CCR, 25 Pa. Code §§287.661-287.666. On January 25, 1997, the beneficial use of CCR regulations, 25 Pa. Code §§287.663 and 287.664 were amended to change the requirements concerning groundwater monitoring, reporting to the Department, CCR beneficial uses, and the amounts of CCR that could be used at active coal mine and abandoned mine sites.

2.5.6. Kentucky

Under Kentucky regulations, CCRs (including fly ash, bottom ash, and scrubber sludge produced by coal-fired electrical generating units) are exempt from regulation as hazardous wastes but are classified as special waste. Excluded from this regulation is boiler slag and residues of refuse-derived fuels such as municipal waste, tires, and solvents. Under Kentucky law, CCRs may be reused under permit by rule regulation (1) as an ingredient in manufacturing a product; (2) as an ingredient in cement, concrete, paint, and plastics; (3) as an antiskid material; (4) as highway base course; (5) as a structural fill; (6) as a blasting grit; (7) as roofing granules; and (8) for disposal in an active mining operation if the mine owner/operator has a mining permit authorizing disposal of special waste. Specific conditions for reuse of CCRs include: (1) the CCR reuse may not create a nuisance; (2) erosion and sediment controls must be undertaken; (3) the CCR reuse must be at least 100 ft from a stream and 300 ft from potable wells, wetlands, or flood plains; (4) the ash must be "nonhazardous"; and (5) the generator must submit an annual report. Mine applications must be specifically authorized under the terms of a permit issued by the Department for Surface Mining, Reclamation and Enforcement.

(KY. REV. STAT. ANN. §224.50-760(1)(a); 401 KY. ADMIN. REGS. 45:010 §(4).)

2.5.7. Tennessee

Under Tennessee law, fly ash, bottom ash, and flue gas emission control waste generated from the combustion of coal or other fossil fuels must be tested for a hazardous waste determination. If determined to be hazardous, certain hazardous waste generator requirements apply. Upon testing confirmation that the material is not hazardous, fly ash, bottom ash, and boiler slag may be reused under permit by rule regulation (1) in engineered structures for a highway overpass, levee, runway, or foundation backfill; and (2) in other proposed beneficial uses approved on a case-by-case basis. Certain restrictions and requirements apply to "permit by rule" uses and proper written notification of the beneficial use must be submitted to the Tennessee Department of Environmental Conservation (TDEC) and approved.

Certain restrictions and requirements apply to "permit by rule" uses. For example, prior written notification of the beneficial use must be submitted to TDEC and approved. The notification must be on forms provided by TDEC. The Permit by rule authorization issued by TDEC must be maintained at the facility.

The project may not be located in wetlands, sinkholes or caves, or in a 100-year flood plain unless certain conditions are met. The potential for releases must be minimized and site access controlled. The project may not cause or contribute to the taking of any threatened or endangered species of plants, fish, or wildlife or result in the destruction or modification of their habitat. Until development is complete, the area must have a barrier to control unauthorized entry. A geologic buffer of at least three feet with a maximum saturated hydraulic conductivity of 1×10^{-6} cm/s must be in place between the fill and the seasonal high groundwater table. A permanent benchmark of known elevation must be installed. All boreholes, piezometers, and abandoned wells within 100 ft of the site must be filled with bentonite in accord with specific requirements.

Within 90 days of completion of the project at least two feet of compacted soil cover must be in place. Final surface grading requirements apply. Dust must be minimized and there must be equipment present at the times coal is received capable of spreading and compacting the coal ash.

(TENN. COMP. R. & REGS. 1200-1-7-0.02(1)(c)(1)(ii).)

2.5.8. Illinois

Under Illinois regulations, fly ash, bottom ash, slag, and flue gas emission control waste generated primarily from the combustion of coal or other fossil fuels are exempt from regulation as hazardous waste. Illinois law specifically authorizes the reuse of CCRs, classified into two different groups: coal combustion waste (CCW) and CCB. CCW reuse is regulated more stringently than CCB. CCW can be classified as CCB under certain conditions and reused, based on the classification, (1) for the extraction and recovery of materials and compounds within the ash; (2) as a raw material in the manufacture of cement and concrete products; (3) for roofing shingles, (4) in plastic products, paints, and metal alloys, (5) in conformance with the specifications and with approval from the Illinois Department of Transportation (IDOT); (6) as an antiskid material, athletic tracks or foot paths (bottom ash); (7) as a lime substitute for soils so long as the CCBs meet the IDOT specifications for agricultural lime as a soil conditioner; (8) in non-IDOT pavement base, pipe bedding, or foundation backfill (bottom ash); (9) as a structural fill when used in an engineered application or combined with cement, sand, or water to produce a controlled-strength material; and (10) for mine subsidence, mine fire control, mine sealing, and mine reclamation (must meet requirements of both the Illinois Environmental Protection Agency (IEPA) and Department of Mines and Minerals). Other CCB applications may be authorized by IEPA.

(35 ILL. ADMIN. CODE §721.104(b)(4) and 415 ILCS 5/3.94 (P.A. 89–93).)

2.5.9. West Virginia

West Virginia regulations adopt by reference the federal regulation that exempts CCRs (including fly ash, bottom ash, slag, and flue gas emission control waste (FGD waste) generated primarily from the combustion of coal) from classification as hazardous waste. Under West Virginia regulations, CCRs may be reused (1) as a material in manufacturing another product or as a substitute for a product or natural resource; (2) for the extraction or recovery of materials and compounds contained within the CCRs; (3) as a stabilization/solidification agent for other wastes if used singly or in combination with other additives or agents to stabilize or solidify another waste product; (4) under the authority of the West Virginia Department of Energy; (5) as pipe bedding or as a composite liner drainage layer; (6) as an antiskid material (bottom ash, boiler slag); (7) as a daily or intermediate cover for certain solid waste facilities; (8) as a construction base for roads or parking lots that have asphalt or concrete wearing surfaces.

(W.VA.REGS. §33-1-5.5.b.4.A-.H.)

2.5.10. Wyoming

Wyoming regulations adopt by reference the federal regulation that exempts CCRs (including fly ash, bottom ash, slag, and flue gas emission control waste generated primarily from the combustion of coal) from classification as hazardous waste. Wyoming law regulates CCBs as an industrial solid waste. Reuse of CCRs is not specifically authorized under Wyoming law. According to the Wyoming Department of Environmental Quality, the Agency is in the process of developing regulatory guidelines for beneficial reuse of solid waste, including CCRs.

Exempt from hazardous waste regulation are fly ash, bottom ash, slag, and flue gas emission control waste generated primarily from the combustion of coal.

(WY ADMIN. CODE HWM CH. 2 §1; 40 CFR 261.4.)

2.5.11. Texas

Texas regulations adopt by reference the federal regulation that exempts CCRs (including fly ash, bottom ash, slag, and flue gas emission control waste generated primarily from the combustion of coal) from classification as hazardous waste. Under Texas regulations, CCRs may be classified as industrial solid wastes. The Texas Natural Resource Conservation

Commission issued CCR reuse guidance, under which CCRs are not subject to classification as a waste and are designated as "coproducts" when used in (1) concrete, concrete products, cement/fly ash blends, precast concrete products, lightweight and concrete aggregate, roller compacted concrete, soil cement, flowable fill, roofing material, insulation material, artificial reefs, and as mineral filler (fly and bottom ash); (2) as a raw feed for concrete manufacture and in masonry (fly ash, bottom ash, and FGD material); (3) in oil well cementing and waste stabilization and solidification (fly ash); (4) as road base when covered by a wear surface; (5) as an unsurfaced road construction material, road surface traction material, and blasting grit (bottom ash); and (6) in wallboard and sheetrock (FGD by-product gypsum material).

(TEXAS ADMIN. CODE 30 §335.2, §335.4, §335.501 *et seq.*; 40 CFR 261.4.)

2.5.12. Ohio

Under Ohio regulations, fly ash, bottom ash, slag, and flue gas emission control waste generated primarily from the combustion of coal or other fossil fuels are exempt from regulation as hazardous waste. Reuse of CCRs is not specifically authorized under Ohio law; however, reuse of "nontoxic" CCRs is authorized under policy documents issued by the Ohio Environmental Protection Agency (OEPA), Division of Water Quality. Nontoxic CCRs may be reused (1) as a raw material in manufacturing a final product; (2) as a stabilization/solidification agent for other wastes that will be disposed; (3) as a part of a composting process; (4) in uses subject to USEPA procurement guidelines; (5) for extraction or recovery of materials and compounds in CCRs; (6) as an antiskid material or road preparation material; (7) for use in mine subsidence stabilization, mine fire control, and mine sealing; (8) as an additive in commercial soil blending operations, where the product will be used for growth of ornamentals (no food crops or grazed land); (9) as daily cover at a landfill; (10) as a structural fill, defined as an engineered use of waste material as a building or equipment supportive base or foundation and does not include valley fills or filling of open pits from coal or industrial mineral mining; (11) as pipe bedding, for uses other than transport of potable water; (12) as a construction material for roads or parking lots (subbase or final cover); and (13) other single beneficial uses of less than 200 tons. Certain guidelines may apply to the above uses.

(OHIO ADMIN. CODE §3745-51-04(B)(4).)

2.6. REGULATORY PROSPECT FOR BENEFICIAL USE

Since the *EDF vs City of Chicago* court decision changed the classification of MWC residues from a nonhazardous classification to hazardous waste, MWC residues have been burdened with a hazardous waste label. The beneficial use of MWC residues, other than as landfill cover, in construction material applications have been stymied due to this hazardous waste categorization.

MWC residue represents about 10% by volume of the trash combusted. Ferrous metals are removed at the facility, leaving a residue that looks a lot like wet cement. WTE residue actually has physical properties similar to construction mixtures such as concrete. After a short time, MWC residue "cures" and resembles concrete (18).

Although technical studies have demonstrated the use of these residues as in road construction and building material, the stigma of hazardous waste and litigation potential has prevented end-users of capitalizing on the technical evidence for MWC residue utilization (19–21).

States overwhelmingly support the beneficial use of CCR in construction applications. Even USEPA's two option proposal (hazardous or non-hazardous) includes the Bevill exemption for beneficial uses of coal ash in which CCRs are used in concrete, cement, wallboard, and other contained applications. The EPA's May 4, 2010 proposal to regulate CCR's as either a Class C Hazardous Waste or a Class D Solid Waste is a political Chinese Menu; although regulating CCBs under either option, the proposal fails to provide regulatory direction. For a material to be classified as hazardous, failure to meet the TCLP must occur. Those, who arbitrarily denote CCBs as hazardous or toxic, do so without substantive TCLP data. Retaining the Bevill exemption, allowing for beneficial use of CCRs, reflects the industry practice of deploying about 40% of the residue stream. Nonetheless should EPA decide to classify CCBs as hazardous or Class C—the potential for end-user liability could reduce by-product utilization. This exemption may translate into EPA regulating CCRs as a special waste (22).

The private sector recognizes severe negative impact of hazardous classification of CCR. In 2008 in the United States, 136 million tons of CCRs were produced, according to a survey by the American Coal Ash Association (ACAA). Almost 16 million tons were used in cement and concrete production. Another 8.5 million tons went into the production of wallboard products. The use of fly ash instead of portland cement, which is an energy-intensive product, avoided 12 million tons of carbon dioxide emissions, says ACAA. "A hazardous waste designation for fly ash would stigmatize its use as an ingredient in concrete" (23).

REFERENCES

1. USEPA; Characterization of Municipal Combustor Ashes and Leachates from Municipal Solid Waste Landfills. Monofills and Codisposal Sites; Publication No. 68–01–7310; September 1987.
2. 40 CFR 261.24.
3. Darcey, S.; "Incinerator Ash Sparks Heated Debate on Toxicity"; World Waste; October 1987, pp 32–36.
4. Kochieisen, C.; "Comparison of Legislation on Municipal Waste Incineration"; Resource Recovery; Vol. 2, No. 3; pp 24–29.
5. Zannes, M.; "Waste-To-Energy Ash Residue"; Integrated Waste Management; September 2003.
6. Coalition on Resource Recovery and the Environment/United States Environmental Protection Agency (CORRE/USEPA) (February 1990); Characterization of Municipal Waste Combustion Ash. Ash Extracts, and Leachates; Contract No. 68–01–7310.
7. Hilbert, J., et al.; "The Garbage Solution Isn't That Simple"; New York Times; February 22, 1987.
8. Florio, J. (Governor of New Jersey); Personal Communications; October 4, 1990.
9. Roukema, M. (Member of Congress); Personal Communications; (January 10, 1991).
10. Kom, J.L.; Huitric, R.L.; "Combined Ash Treatment for the Commerce Refuse-to-Energy Facility"; presented at ASTSWMO Nat'l Solid Waste Forum; Portland, OR; July 20–22, 1992.
11. Goodwin, R.W.; "Utilization Scenarios For High-Volume Residues From Coal-Fired Power Plants and Resource Recovery Facilities"; Proceedings of 4th International for Power Generating Industry (Power-Gen '91); Tampa Convention Center; Tampa, FL; December 4–6, 1991.
12. USEPA Soild Waste and Emergency Response; EPA 530-K-02–0191; "RCRA, Superfund & EPA RCRA Call Center Training Module"; October 2001, pp 14–15.
13. Worobel, R.; Leo, K.; Gorrie, J.; Thur de Koos, P.; Bruner, M.; "Beneficial Reuse of Municipal Waste-to-Energy Ash as a Landfill Construction Material"; ASME NAWTEC Paper No. 11-1678; 11th North American Waste to Energy Conference; April 2003.
14. NUS Corp.; Characterization of Municipal Waste Combustion Ash, Ash Extracts, and Leachates; USEPA Contract No. 68–01–7310; February 1990.
15. Goodwin, R.W.; Residues from Waste-to-Energy Systems; Comments Submitted to USEPA Pursuant to Proposed Amendment Subtitle C of RCRA (40 CFR Parts 261, 271 and 302); 7/31/86.
16. California Dept. of Health Services (CDHS); "Classification of Stanislaus Waste Energy Company Facility Ash"; 2/8/90.
17. New York State Department of Environmental Conservation, Division of Solid Waste, Bureau of Resource Recovery; Ash Residue Characterization Project; March 1992.
18. Ormsby, W.C.; Federal Highway Administration; "Paving with Municipal Incinerator Residue"; Proceedings of the First International Conference on Municipal Solid Waste Combustor Ash Utilization, at 49; October 1988.
19. Musselman, C.N., et al.; "Utilizing Waste-to-Energy Bottom Ash as an Aggregate Substitute in Asphalt Paving"; Proceedings of the Eighth International Conference on Municipal Solid Waste Combustor Ash Utilization, at 59; November 1995.
20. Jones, C.M., et al.; "Utilization of Ash from Municipal Solid Waste Combustion"; Final Report, Phase I, NREL Subcontract N XAR-3-1322 at 13; 1994.
21. Roethal, F.J.; Breslin, V.T.; "Stony Brook's MSW Combustor Ash Demonstration Programs"; Proceedings of the Third International Conference on Municipal Solid Waste Combustor Ash Utilization, at 237; November 1990.
22. Peltier, R.; "My Top 10 Predictions for 2013"; Power, January 2013, p 6.

23. O'Hare, A., Vice President of Regulatory Affairs, Portland Cement Association; Personal Communication; 4/7/2010.
24. Korsun, E.A.; Heck, H.H.; "Sources and Fates of Lead and Cadmium in Municipal Solid Wastes"; Journal of the Air and Waste Management Association; Vol. 40, No. 9, September 1990; pp 1220–1226.
25. Francis, C.W.; White, G.H.; "Leaching of Toxic Metals from Incinerator Ashes"; Journal Water Pollution Control Federation; Vol. 59; (11), 1987; pp 979–986.
26. Wiles, C.C.; "The U.S. EPA Program for Evaluation of Treatment and Utilization Technologies for Municipal Waste Combustion Residues"; Municipal Waste Combustion; Air & Waste Management Assoc.; Tampa, FL; April 15–19, 1991.
27. Hansen, P.L.; "Ash Stabilization, a Case Study"; Hazardous Material Control; (1/2), 1991; pp 13–17.

CHAPTER *3*

Regulatory Testing

Contents

3.1. INTRODUCTION

This chapter discusses how the pozzolanic-like behavior of municipal waste combustor (MWC) residues should be recognized and incorporated in regulatory testing procedures. Regulators have focused on laboratory leachate tests to establish the classification of ash from municipal solid waste (MSW) incinerators (1). Such attention raises several issues, which range from the

Combustion Ash Residue Management
http://dx.doi.org/10.1016/B978-0-12-420038-8.00003-3

applicability of previous federal policies, to the manner of sampling, and to the type of test to apply to characterize the ash (2–4). A detailed discussion of such issues is not the intent of this chapter; they are addressed in terms of actual practice. The state regulatory agencies issue environmental permits for the construction/operation of resource recovery facilities. As a part of this permit, the ash is typically tested in accordance with the USEPA's Extraction Procedure Method (5,6) and recently supplanted with the Toxic Characteristic Leachate Procedure (Federal Register, September 25, 1990). Due to the wide variation of the MSW feedstock, the ash should be composite sampled. Even under such a scenario, laboratory tests have shown that MWC ash exceeds regulatory requirements.

3.2. REGULATORY HEAVY METAL LABORATORY LEACHATE TEST RESULTS (OLDER SYSTEMS)

The ash generated from MWC systems is composed primarily of bottom ash and fly ash. Several laboratory studies of operating MWC systems have shown that, possibly due to its smaller size, fly ash exceeds the EP (Extraction Procedure) Toxicity regulatory limit in terms of heavy metals (predominantly lead (Pb) and cadmium (Cd)) (4,6) and exhibits higher concentrations of organic contaminants (i.e. polychlorinated biphenyls (PCBs), polychlorinated dibenzo-p-dioxins (PCDDs), polychlorinated dibenzofurans (PCDFs)) (4). Such results, however, are based on older MWC systems, which are not equipped with acid gas cleaning (i.e. dry scrubbers). This chapter discusses the theoretical and actual effect of the lime (made available from the dry scrubbers) upon both properties and classification of ashes from MWC systems. The reaction of this available lime with the other mineral constituents of the ash forms a concrete-like substance or pozzolan rendering the resultant product both nonhazardous and suitable for by-product utilization.

Under the Resource Conservation and Recovery Act, Subtitle C, the heavy metal content of a material indicates, in part, its hazardous classification. A waste's heavy metal content is analyzed according to a predictive lab leachate test, e.g. EP Toxicity Procedure (6). Applying this analytical procedure is critical in resolving the hazardous waste classification of MWC residues.

Taylor (7) conducted elemental heavy metal analyses of MWC ash heavy metal as a function of collection efficiency. This Maryland-based work reponed arsenic concentrations as high as 1619 micrograms per gram (μg/g).

This manner of reponing, i.e. elemental analysis, could distort the interpretation of the data; implying a hazardous classification of the residue. Rademaker and Young (8), utilizing the appropriate analytical procedure, provide a legal and accurate assessment of MWC residue hazardous potential. Although these results indicate that the Nashville facility residues could be deemed nonhazardous, in some instances the heavy metal content of another installation could exceed the maximum allowable concentrations and would be classified a hazardous waste. Tennessee's Sumner facility exhibited excessive cadmium and lead (32.0 and 14.5 mg/l, respectively) levels in the fly ash and excessive lead concentrations (6.98 mg/l) in the bottom and fly ash mixture. For the three facilities sited only particulate removal was implemented. The present determination of the heavy metal content of these residues should consider dry flue-gas desulfurization (FGD) or acid gas treatment to appropriately consider and predict the heavy metal content of residues from such facilities.

3.3. POZZOLANIC CHARACTERISTIC ANALOGIES— CHEMICAL COMPOSITION

Consider the MWC ash analysis of a proposed waste-to-energy facility in New Jersey as shown in Table 3.1.

Coal-fired power plant experience forms the basis for suggesting that the lime and fly ash will form a pozzolan capable of encapsulating the heavy metals contained in the MWC residues (9). Lime (CaO) in the presence of silica (SiO_2), alumina (Al_2O_3), and calcium sulfate ($CaSO_4$) form sulfo-alumina hydrates (ettringites) and calcium silica hydrate (tobermorite) as represented by the following:

$$\text{Ettringite: } 3CaO \cdot Al_2O_3 \cdot 3CaSO_4 \cdot 28\text{--}32H_2O$$

$$\text{Tobermorite: } CaO \cdot SiO_2 \cdot nH_2O$$

Table 3.1 Constituents of MWC Residue

Constituent	Percent by Weight
Fly ash	49
Calcium chloride	12
Calcium sulfate	14
Lime	25

Table 3.2 Comparisons of Utility FGD Waste to Residue and APC Waste

Component	Utility DFGD Waste	MWC Residue/APC Waste
SiO_2	28	24
Al_2O_3	14	6
Fe_2O_3	5	3
CaO	24	37
$CaCl_2$	4	12

Note: Mineral components shown as wt%.

The power plant analogy for inherent pozzolanic behavior can be demonstrated by considering a comparison of utility (power plant) waste to MWC residue, as shown in Table 3.2.

Utility waste from dry FGD systems exhibits both pozzolanic behavior and heavy metal analysis, below maximum concentrations (10). Comparison of the silica, alumina, and lime content suggest that similar reactions should occur with MWC residues. The considerably higher lime content of MWC residues is due to the higher stoichiometric requirements of dry acid gas systems.

3.3.1. Discussion of Significant Constituents

A discussion/comparison of the significant constituents within the two types of waste materials provides a basis for suggesting (1) disposal acceptability and (2) potential for by-product utilization.

3.3.1.1. Silica (SiO₂)

SiO_2 forms calcium silica hydrates (tobermorite), dicalcium silicate (C_2S), and tricalcium silicate (C_3S), which are the main strength-contributing components of the pozzolanic behavior. Given similar SiO_2 contents the resultant structural-yielding reaction for the MWC residue should be satisfactory.

3.3.1.2. Lime (CaO)

The lime (CaO) content directly contributes to the strength and permeability of the end-product. Tobermorite is chemically represented as $CaO \cdot SiO_2 \cdot nH_2O$. The higher lime (CaO) content of the MWC residues suggests attaining analogous geotechnical properties.

3.3.1.3. Alumina (Al₂O₃) and Iron Oxide (Fe₂O₃)

Al_2O_3 behaves similar to SiO_2 forming aluminates and the end-product, ettringite ($3CaO \cdot Al_2O_3 \cdot 3CaSO_4$). Fe_2O_3 forms the ferrite phase, which reacts with the aluminates and silicates. High CaO values in the MWC residue suggest a compensatory effect due to its lower Al_2O_3 and Fe_2O_3 content.

3.3.1.4. Chlorides (CaCl$_2$)

$CaCl_2$ is important to concrete chemistry. At contents >1.5% a morphological change occurs which increases the end-product's porosity and permeability (11). Strengths at these elevated chloride levels, however, are relatively unaffected. The significantly higher chloride content in the MWC ash may be mitigated by the presence of higher CaO.

While resolution of the heavy metal issues can be reasonably expected in view of pozzolanic reactions, the organic content of these residues must also be evaluated.

3.3.2. Organic Content of MWC Residues

The issues of organic (e.g. dioxin and furan) destruction and removal can be considered by implementing both high operating incineration temperatures and dry acid gas removal with fine particulate removal.

A low probability exists for dioxin/furan formation in MWC incinerators operating at or in excess of 1200 K (12). Even at incinerator operating temperatures equal to or greater than 1000 °C (1800 °F), 99.99% destruction of PCB's, dioxins, and furans have been achieved, provided that sufficient oxygen and mixing are implemented within the incinerator design (13). Even though virtually total organic destruction is expected, given these incinerator operating conditions, the dry acid gas system provides an additional margin of safety.

Typically a dry acid gas system consists of a spray dryer absorber (SDA) and a baghouse (BH). When this system operates at temperatures equal to or less than 300 °F, the dioxins and furans will condense onto the fine particulates (i.e. <2 μm or 0.02 gr/scfm (grains per standard cubic feet per minute)) (13).

Consider the data (Table 3.3) from a European MSW facility equipped with SDA and BH (14):

The increased capture of these organics in the BH catch will add dioxin and furans to the residue composition. The pozzolanic reaction has not been demonstrated for potentially encapsulating these organic compounds. Additional research and development efforts are required to address the potential hazardous characterization of MWC residues containing dioxins and furans.

Table 3.3 Effectiveness of SDA and BH to Remove Organics

	SDA-In (ng/Nm³)	BH-Out (ng/Nm³)	% Removal
Total dioxin (PCDD) High temp.	146	23	84
Low temp.	235	<1.15	>99.5
Total furan (PCDF) High temp.	237	24	90
Low temp.	531	<1.4	>99.7

Note: ng/Nm^3 = nanograms per normal cubic meter.

3.3.2.1. Subsequent Dioxin/Furan Studies

This section cites results from two studies of dioxin and furan content in and leachate from MWC residues, i.e. USEPA/CORRE and Sumner County TN, respectively. Evaluation of their results resolves the issue of these organic compounds posing potential health and/or environmental risk. Based upon such *de minimis* regulatory sponsored and reviewed results, no further discussion of organic constituents is necessary.

In their study of MWC residues from five operating waste-to-energy facilities, USEPA/CORRE analyzed five ash samples from each of the facilities for PCDDs/PCDFs. According to Ref. (15), the procedure described by the reference's cited "Interim Procedures for Estimating Risk Associated with Exposures to Mixtures of chlorinated dibenzo–p–dioxins/ dibenzofurans (CDDs/CDFs)" were used to develop a toxicity equivalent. This equivalent was compared to the Center for Disease Control's recommended upper limit of one part per billion (ppb). The total equivalent for each ash sample was below 1 ppb.

Resource Authority in Sumner County, TN analyzed its 11/26-27/90 ash monofill leachate for total chlorinated TCDDs. Their results indicate TCDDs <0.01 parts per trillion.

3.4. POZZOLANIC EFFECT—OPERATING RESULTS

The previous analogous discussion of the potential pozzolanic behavior of MWC ash (due to its favorable mineral composition) and of the relatively high lime content of MWC ash (due to higher stoichiometrics, i.e. no recycle) provides the theoretical basis for considering heavy metal reduction and enhancement of geotechnical properties. Achieving such behavior depends upon proper sample preparation.

The power plant analogy affords a database, which justifies applying the principles of maximizing geotechnical properties via optimization of moisture content and lime solubilization. When the percent solids of lignite-fired dry FGD waste was reduced from 80 to 60 wt%, the unconfined compressive strength (UCS) increased from 3.4–6.7 ton/ft^2 (TSF) to 57.8–70.8 TSF; showing the impact of lime solubilization. The effect of moisture content optimization was demonstrated by slightly increasing the solids content of dry FGD waste (from low alkaline coal) from 75 to 80 wt%; achieving UCS increases from 2.14 to 4.06 TSF. The permeabilities of these materials ranged from 2.5×10^{-5} to 6.2×10^{-7} cm/s (16).

3.4.1. Geotechnical Properties: MWC Residues

This power plant waste site management technology has been applied to MWC ash. A typical compaction procedure (ASTM D-698) was conducted on MWC ash representing 240 ton/day (TPD), at a waste-to-energy facility equipped with dry lime injection. The compliance stoichiometric ratio of this system has been reported as 3.4 (17); implying considerable unreacted lime in the ash. Optimal moisture occurred at 20.5%, achieving a maximum dry density of 106 lb/ft^3 (dry). The permeability associated with this MWC ash reached 1.6×10^{-5} cm/s, lying within typical permeabilities of pozzolanically behaving residues (18,19).

MWC's ashes' pozzolanic behavior, not only exhibits low permeability, but considerably reduces EP Toxicity of heavy metal concentrations.

3.4.2. Heavy Metal Leachate Reduction

Only lead and cadmium are reported, since they are the most problematic heavy metals. Table 3.4 depicts lead and cadmium EP Toxicity analysis (20) of ash obtained from an operating waste-to-energy facility (700 TPD) located in New England. These results reflect the range of values obtained from samples taken during the spring of 1987 (April–June). This period included the start-up/shake-down of the facility. When the dry scrubber was not operating, due to its coupling to the incinerator's full load conditions, the lead (22.0–7.10 mg/l) and cadmium (1.5–2.4 mg/l) exceeded regulatory limitations. Upon operating the dry scrubber at design conditions, the lead (0.40–0.50 mg/l) and cadmium (0.20–0.53 mg/l) dropped to acceptable levels.

Typical bottom ash EP Toxicity lead (3.2 mg/l) and cadmium (0.35 mg/l) are also shown on Table 3.4. Extrapolating from the averages of the Pb and

Table 3.4 Heavy Metal EP Toxicity Analysis: Intermittently Operating Dry Lime Scrubber

Element	Without Dry Scrubber	With Dry Scrubber	Maximum Concentration
Cadmium	2.4–1.5	0.20–0.53	1.0
Lead	22.0–7.10	0.40–0.50	5.0
Typical Lead and Cadmium EP Toxicity Analysis Ash			
Element	Bottom Ash		Maximum Concentration
Lead	3.20		5.0
Cadmium	0.35		1.0

Note: All concentrations expressed in mg/l.

Cd ranges yields fly ash levels of Pb = 31.58 mg/l and Cd = 4.35 mg/l, consistent with other investigators (21). The reduction of the lead and cadmium levels of the resultant ash by one to two orders of magnitude cannot be attributed to just dilution by bottom ash addition; rather the pozzolanic reaction reduced the resultant heavy metals from leaching under the EP Toxicity Test.

3.5. SAMPLE PREPARATION REFLECTING INHERENT POZZOLANIC BEHAVIOR

Due to the unique chemistry of waste-to-energy residues (for systems equipped with dry acid gas treatment) a pozzolanic reaction could occur. Such behavior should be recognized by the USEPA and state regulatory agencies in allowing appropriate sample preparation of this material prior to application of either the USEPA's previous (EP Toxicity) or their recently substituted Toxicity Characteristic Leaching Procedure.

Researchers have shown that contaminant (e.g. heavy metal) encapsulation will occur via the pozzolanic reaction, provided that the optimal moisture content is obtained to achieve both free lime solubilization and maximum compacted density. By establishing a sample preparation technique based on soil testing procedures (to achieve optimal geotechnical properties and to promote lime solubilization, as applied in a similar manner as residues from coal-fired power plants) leachate testing of these samples could occur after curing (i.e. 3–28 days). Thus the pozzolanic behavior is obtained and any contaminant encapsulation will be realized.

3.5.1. Sample Preparation Procedures

Since the residues exhibit soil-like behavior, standard American Society for Testing Materials (ASTM) testing procedures have been modified to account for their fine-particle consistency which induces thixotropic characteristics.

Typically, the residues will exit the waste-to-energy systems at a solids content between 75 and 85% to prevent fugitive dusting and to prevent premature setup, i.e. pozzolanic behavior in the conveying vehicle, e.g. truck or container. In the field the moisture content of the material will be modified to achieve both maximum geotechnical properties and to solubilize the lime.

3.5.1.1. Temporary Curing: Nonsoil-Like Consistency

The material must be able to form a cylinder for testing purposes. Hence, testing fresh material should be limited to a soil-like, nonthixotropic consistency. Additional testing, e.g. leachate testing, cannot be performed when a material has a solids content exhibiting noncompactible conditions. Additional testing, therefore, should be applied to a final product material having a solids content range exhibiting a compactable soil-like characteristic.

At a solids content of noncompactible characteristic, the material must be cured until a soil-like, nonthixotropic consistency is achieved. Care should be taken to form the cylinder before considerable pozzolanic action occurs, preventing disruption or reactants. Such a condition is similar to remolding an in-situ sample.

3.5.1.2. Compaction and Specimen Preparation

Compaction can only be achieved if material exhibits soil-like properties, i.e. nonthixotropic. In some cases, temporary curing will be required to achieve a compactable, soil-like consistency of material. A unit compactive effort should be applied in accordance with ASTM Procedure D 1557, Method A.

3.5.1.3. Modification and Supplementary Information for Compaction and Moisture Density

The compactive effort should be applied in accordance with ASTM Procedure D 1557, Method A. This procedure requires compacting the ash and APC waste in a mold using a 4.54 kg (10 lb) rammer and a 0.457 m (18 in) drop. The mold is a 0.102 m (4 in) diameter cylinder into which approximately three equal layers of MWC residue (totaling approximately 0.127 m (5 in) of compacted depth) are compacted by 25 uniformly distributed blows from a rammer. The moisture content is measured by ASTM D 2216, and the density of the compacted material is determined by measuring the volume and weight.

3.5.1.4. General Comments
Material must pass either a No. 4 sieve (4.75 mm (0.187 in)) or 3/4-in sieve. This test determines the dry unit weight in pounds per cubic foot of compacted ash.

3.5.1.5. Test Material Restrictions
Particle size determinations should be conducted on material before temporary curing and/or compaction in cylinder molds to determine particle size and if a larger (6 in) mold is required. "Compaction reduces air voids in soil, with little or no reduction in water content" (22).

Laboratory compaction tests can correct for greater than 3/4-in particle content by identifying fractions greater than 25% weight. The California Bearing Ratio mold can be used for this type of material (23).

3.5.1.6. Soil Particle Susceptible to Crushing
If the sample contains granular material similar to soft limestone, sandstone, or other minerals likely to be broken down by the action of the rammer, separate batches should be prepared for compaction at each moisture content. ASTM D 1557 should be amended as discussed in the next section.

3.5.1.7. Prepare Batches for Test
From the sieved sample take five or more representative samples each of about 2.5 kg. Mix each batch with a different amount of water to give a suitable range of moisture contents. The moisture content of the first batch could be as suggested below (Section 3.5.1.8). Additional increments of water could be added to the other batches as suggested below (Section 3.5.1.9). The range of moisture contents should be as stated below (Section 3.5.1.10).

Thorough mixing of the water is essential. With fine-particle clay-like materials each batch should be sealed and stored overnight before proceeding with the test.

3.5.1.8. Prepare Test Sample
Riffle the sieved material so as to obtain a representative sample of about 5 kg. Add a suitable amount of water, and mix thoroughly. The amount of water depends on the particle size and distribution and is analogous to typical soil types.

Sandy and gravelly soils: 4–6% (200–300 ml of water to 5 kg of material). Cohesive soils: about 8–10% below plastic limit, i.e. (PL-10) to (PL-8). For example, an ash of PL = 20 should be made up to a moisture content of 10–12% (add 500–600 ml of water to 5 kg of material). Thorough mixing of the water is essential. With fine particle clay-like materials, the mixed sample should be stored overnight in a sealed container before proceeding with the test.

3.5.1.9. Breakup and Remix

Breakup the material on the tray, by rubbing through a 20 mm-sieve if necessary, and mix with the remainder of the prepared sample. Add an increment of water, approximately as described below.

Material whose particle size and distribution is similar to sandy and gravelly soils: 1–2% (50–100 ml of water to 5 kg of soil). Material whose particle size and distribution is similar to cohesive soils: 2–4% (100–200 ml of water to 5 kg of material). Mix in the water thoroughly.

3.5.1.10. Repeat with Added Water

Repeat standard test procedures for each increment of water added, so that at least five compactions are made. The range of moisture contents should be such that the optimum moisture content (at which the dry density is maximum) is within that range. If necessary to define the optimum value clearly, carry out one or more additional tests at suitable moisture contents. Keep a running plot of density against moisture content so as to see when the optimum condition has been passed.

Above a certain moisture content the material may be extremely difficult to compact. For instance, a granular-like ash may then contain excessive free water, or a clay-like ash may be very soft and sticky. In either event the optimum condition has been passed and there is no point in proceeding further.

3.5.1.11. Final Curing of Optimally Compacted Sample Prior to Additional Testing

Upon achieving a soil-like, compactable state, the specimen can be prepared for additional testing, e.g. leachate testing. The specimen will be compacted at the unit compactive effort to attain optimal geotechnical properties, i.e. maximum compacted density at optimal moisture content (see Figure 4.1 of Chapter 4).

In general, the material will be cured for 28 days in a humidity chamber in accordance with ASTM C 511. Prior to additional testing, these specimens will be cured at 60 °C (140 °F) for 28 days.

By permitting such sample preparation techniques, reflective of inherent pozzolanic behavior, the heavy metals leachability will be minimized. Such testing procedures, designed to mimic field behavior, recognize MWC residues' self-hardening properties—an intrinsic consideration for the application of engineering principles for landfilling (Chapter 4) and implementing a by-product utilization scenario (Chapter 5). Avoiding the hazardous classification of residues from waste-to-energy systems, permits end-users (e.g. construction contractors) to transport these materials from the waste-to-energy system to the construction site without unnecessary liability. Nonrecognition of these properties and procedures, by regulators, imposes technically unjustified barriers to reuse.

REFERENCES

1. Chemical Engineering; 4/14/86; p 27.
2. Chemical & Engineering News; 4/14/86; p 21.
3. Engineering Times; April 1986; p 3.
4. Newsday; 10/11/85; p 11.
5. Princeton Aqua Science; Resco Plant Residue Ash Characterization Study; 4/12/76.
6. Federal Register; 7/1/81; p 355.
7. Taylor, D.R.; Environmental Science & Technology; Vol. 16, No. 3, 1982.
8. Rademaker, A.D.; Young, J.C.; ASCE Journal of the Energy Division; Vol. 107, No. EY1, May 1981.
9. Goodwin, R.W.; "Chemical Treatment of Utility and Industrial Waste"; ASCE National Conf. Environmental Engineering; July 14–18, 1982; Minneapolis, MN.
10. Goodwin, R.W.; "Air Pollution Cleaning Wastes: Dry vs. Wet"; ASCE Journal of Energy Engineering; Vol. 109, No. 3, September 1983.
11. Rixom, N.R.; Chemical Admixtures for Concrete; E&F N. Spon Ltd; London; 1978.
12. Shaub, T.; Tsang, Y.; Environmental Science & Technology; Vol. 17, No. 12, 1983.
13. Lauber, J.; Personal Communications; NYSDEC; 2/86.
14. Nielsen, J.; Personal Communications; Niro (Sweden); 1983.
15. CORRE/USEPA; Characterization of Municipal Waste Combustion Ash, Ash Extracts, and Leachates; Contract No. 68–01–7310; February 1990.
16. Goodwin, R.W.; "Waste Treatment and Disposal Aspects: Combustion and Air Pollution Control Processes"; Journal of the Air Pollution Control Association; Vol. 31, No. 7, July 1981, pp 744–747.
17. Foster, J.T., et al; "Design and Start-Up of Dry Sorbent Injection System for Solid Particulate an Acid Gas Control on a Municipal Refuse-Fired Incinerator"; presented at the Industrial Gas Cleaning Institute Forum '87; Arlington, VA; September 22–23, 1987.
18. Goodwin, R.W., et al; "Properties of Stabilized FGD Sludge"; presented at APCA Conf.; 6/25–30/78, Houston, TX.
19. Goodwin, R.W.; "Land Disposal: Fly Ash and Sludge"; presented at American Power Conference; Chicago, Illinois, April 1980.
20. USEPA; SW 846 – Test Methods for Evaluating Solid Waste: Physical/Chemical Methods; 3rd Edition; September 1986.
21. Repa, E.; "The Confusion and Questions about Ash"; Waste Age; September 1987, pp 89–92.
22. Head, K.H.; Manual of Soil Laboratory Testing; John Wiley & Sons; NY; 1980.
23. Lambe, T.W.; Soil Testing for Engineers; John Wiley & Sons; NY; 1951.

CHAPTER 4

Disposal Considerations

Contents

4.1. INTRODUCTION

The waste-to-energy facility's primary concern regarding municipal waste combustor (MWC) ash involves yielding a transportable material. This ash, however, upon exiting the mechanical conveying system, awaiting transport to a landfill, is subject to regulatory testing and to proper disposal site management. This chapter offers data demonstrating the pozzolanic behavior of MWC ash and incorporates such results into ash management principles to ensure setup. Regulatory agencies should review this technical information and allow the application of basic civil engineering and concrete chemistry to be reflected within sample preparation procedures (Chapter 3).

Combustion Ash Residue Management
http://dx.doi.org/10.1016/B978-0-12-420038-8.00004-5
43

Typically the ash from older resource recovery systems has been placed on top of the existing sanitary landfill. This practice, termed codisposal, however, has been replaced by separate ash disposal or monodisposal. An example of such monodisposal or monofill is Warren County, NJ facility. Independent of the type of ash landfill design, the ash should be placed or managed according to sound engineering principles.

Typically, the residues should exit the waste-to-energy systems at a solids content between 75–85% to prevent fugitive dusting and to prevent premature setup, i.e. pozzolanic behavior in the conveying vehicle, truck, or container (per Figure 1.2). In the field, the moisture content of the material should be modified to achieve both, maximum geotechnical properties and to solubilize the lime.

4.2. OPTIMAL DISPOSAL SITE MANAGEMENT FOR MWC RESIDUES

To ensure achieving maximum geotechnical properties, a civil engineering site management approach is depicted by Figures 4.1 and 4.2. Figure 4.1 depicts the traditional moisture–density relationship, while Figure 4.2 shows the relationship of final solids content upon geotechnical property as a function of free, available lime.

Work performed on coal-fired power plant, alkaline, fly ash/APC residues has confirmed that high strengths and low permeabilities are achieved (1,2). Since the residues from waste-to-energy systems will be land-disposed in the same manner as coal-fired waste, the testing procedure should consider the chemical reaction of the lime with the ash constituents prior to leachate testing.

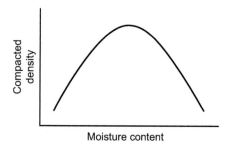

Figure 4.1 *Typical Compaction Curve.*

4.2.1. Compaction

MWC ash possesses the potential for pozzolanic behavior due to the presence of high free lime content. Recognizing this potential pozzolanic reaction is critical (1) to implement any contemplated residue testing requirement and (2) to incorporate into the residue disposal site management practices.

The author has discussed such technology in terms of a graphical-mathematical approach to flue-gas desulfurization (FGD) sludge and power plant ash (3). Inherent to this approach are (1) achieving optimal moisture content and (2) solubilization of free, available lime. Achieving these objectives assures pozzolanic behavior of residues. Since the purpose of laboratory testing is to simulate/predict the residue behavior in the field; incorporating these principles during MWC ash sample preparation (prior to testing) is reasonable and appropriate. A detailed discussion of such sample preparation procedures has been presented in Chapter 2 (4); a summarized general approach is listed as follows:

- temporary curing: nonsoil-like consistency;
- compaction and specimen preparation (lime solubilization);
- curing of optimally compacted samples.

4.2.2. Achieving the Inherent Pozzolanic Behavior

The previous discussion of the potential pozzolanic behavior of MWC ash (due to its favorable mineral composition) and of the relatively high lime content of MWC ash (due to higher stoichiometries, i.e. no recycle) provides the theoretical basis for considering heavy metal leachate reduction and enhancement of geotechnical properties. Achieving such behavior depends upon solubilizing the free, available lime and attaining optimal compaction.

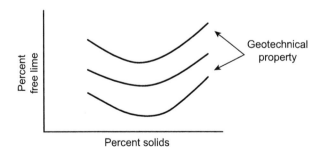

Figure 4.2 *Optimizing Lime/Water Content vs Final Property.*

Table 4.1 Effect of Percent Solids/Water on Permeability

% Solids	Permeability (cm/s)	
	After 120 h	After 28 days
75	2.5×10^{-7}	1.31×10^{-8}
80	2.3×10^{-5}	1.02×10^{-8}

Laboratory work has revealed an inverse trend relationship between (1) percent solids/water of solubilization and (2) mean particle size. To achieve a geotechnical property (i.e. strength, permeability), reflecting setup conditions, more water of solubilization was required for a recipe with finer particle size distribution. A possible explanation for this relationship is that the smaller sized particles exhibit a greater surface area thus requiring more water of solubilization within the voids to promote the pozzolanic or setup behavior.

4.2.3. Effect of Water of Solubilization on Setup Time

Not only does the introduction of additional water of solubilization facilitate attaining pozzolanic behavior but optimizing the percent water of solubilization reduces the setup time. When 15% Portland Cement (PC) was added to combined ash (CA) (i.e. bottom and fly ash), the permeabilities and curing time were determined (Table 4.1).

Achieving the significantly lower permeabilities (i.e. two orders of magnitude reduction) at the early cure time (120 h), when more water was present in the sample, suggests that adding more water of solubilization accelerates the pozzolanic behavior. An aqueous phase is more quickly established for the pozzolanic constituents to react. Since 28 day permeabilities were essentially the same for both percent solids samples, the final pozzolanic property was achieved for these ashes of equivalent composition. With two ashes of equivalent pozzolanic composition, the setup reaction time is increased (i.e. lower permeability) when more water is available for solubilization of reactive constituents.

4.2.4. Free Available Lime and Curing Time

As previously discussed, when the CA contains excess, free available lime (due to the high stoichiometries associated with acid gas cleaning systems), pozzolanic behavior occurs. The results of Table 4.2 demonstrate the effect upon permeability when 3–9% lime (CaO) was added to these ashes with the optimum compaction, percent water of solubilization and curing.

Table 4.2 Effect of Lime, Dosage, and Curing on Permeability

Ash	% CaO Additive	Curing Time (days)	Permeability (cm/s)
CA	3	0	1.0×10^{-6}
CA	3	7–14	8.1×10^{-7}
CA	6	0	2.2×10^{-6}
CA	6	7–14	3.5×10^{-7}
CA	9	0	2.2×10^{-6}
CA	9	7–14	8.0×10^{-7}

Achieving permeabilities between 10^{-6} and 10^{-7} cm/s is significant, since such permeabilities are used for landfill liners. These results, therefore, suggest that for a given ash/additive menu a finite percent water of solubilization occurs which achieves the pozzolanic reaction within the quickest time and to the highest attainable geotechnical property.

The lowest permeability, achieved at 6% CaO, further supports the optimization concept of adding a specific amount of solubilization water for a specific blend of materials. By optimizing percent water of solubilization to a given ash/additive particle size distribution, a minimal additive dosage can be established. Full-scale correlation of additive feed rate to online monitoring of particle size distribution could minimize additive chemical costs while achieving the desired geotechnical property.

4.3. COMPACTION REDUCES ASH DISPOSAL REQUIREMENTS

The MWC ashes investigated in the laboratory program were generated from resource recovery facility I, equipped with a lime-based APC system. Unlike facility I, the ash studied in the field program was generated at an older facility II, not equipped with a lime-based APC system.

Table 4.3 depicts the particle size distribution of raw bone ash (BA) and combined ash (CA) from facility I. BA (I) reflects a granular material of mostly gravel consistency. CA (I) represents a finer material, having almost equal gravel and sand consistencies. These ashes were compacted in accordance with the modified Proctor test (ASTM D–1557). The results of this testing are shown on Table 4.3. Table 4.3 also depicts the particle size distribution and Proctor testing of CA (II). This distribution, twice as much gravel consistency as sand, is bracketed by BA (I) and CA (I) distributions and reflects proprietary mechanical processing.

Table 4.3 Comparison of Size Distribution and Compacted Density (6)

Test Patch Placement	% Moisture[a]	Density (lb/cu ft)[a]	
		Wet	Dry
Pre-compaction	18.15	123.52	104.58
Post-compaction	19.00	134.82	113.31

Ash	D_{max}	D_{mean}	D_{min}	Compacted DD
BA (I)	30	7.2	0.04	92.1
CA (I)	20	3.0	0.04	74.9
CA (II)	12	4.2	0.04	113.31

D = diameter (mm); DD = dry density (lb/cu ft).
[a]In-place test patch measurements.

All ashes were compacted at a nonoptimal moisture content to facilitate achieving the desired percent water of solubilization. Ashes from facility I represent modified Proctor compactions, while CA (II)'s compacted densities and moisture contents were measured in the field, with a nuclear-density meter. These measurements were obtained during 10 hours of the 1-day placement activities. The in-situ moisture content results include a field-correcting factor, whose error was less than ±1%. Table 4.3 provides a comparison of particle size distribution and compacted dry density for all three ashes. The compactive effort of the modified Proctor test is suitable to simulate heavy field equipment used in airfield construction (5). The compacted density of CA (II) was achieved by compaction with a 20-ton vibratory roller. Therefore, the compactive effort in the laboratory and field are comparable. The highest density, achieved for CA (II), reflects its proprietary mechanical processing. Table 4.3 also reports the in-place density and moisture content of the CA (II) before and after compaction. The ash's wet density of 134.82 pounds per cubic foot (lb/cu ft) represents the unit weight of the solids and water as compacted in-place. In-place compaction and solubilization water addition exemplify the application of engineering principles to achieve enhanced geotechnical properties.

4.3.1. Volumetric Reduction

The application of an engineering approach to ash disposal practice attains higher densities and lower volumetric requirements. The significance of obtaining compacted dry density of 113.31 lb/cu ft for this mechanically processed and well compacted ash, can be appreciated by considering typical ash densities.

Typically, MWC ash landfills have been sized on the basis of a wet density $= 74\,lb/cu\,ft$ for % water content $= 15–25\%$. By assuming 20% water, the typical dry density for landfill sizing $= 59.2\,lb/cu\,ft$. The following derivation represents volumetric reduction in terms of initial and final densities:

$$\text{Change in volume} = (V_o - V_f)/V_o \qquad (4.1)$$

where $V_o =$ initial volume; $V_f =$ final volume; $V_o > V_f$.

Since $DD_o =$ initial dry density $= W_o/V_o$ (where $W_o =$ initial weight) and $DD_f =$ final dry density $= W_f/V_f$ (where $W_f =$ final weight) and $W_o = W_f$ (i.e. dry solids remain the same), then

$$\% \text{ Volumetric reduction} = [1 - (DD_o/DD_f)] \times 100 \qquad (4.2)$$

Both mechanical processing and field compaction significantly reduce volumetric ash monofill requirements. By applying Eqn (4.2), achieving a 48% volumetric reduction over typical ash landfill practices is determined. Field compaction amounted to an 8% volumetric reduction. The remaining 40% volumetric reduction was due to proprietary mechanical processing, yielding an improved solids distribution. The shape of a particle size distribution affects the nesting of the different sized particles. A material, having uniform distribution of large, medium, and fine sized particles, should permit more solids occupancy and less voids space for a given volume of material (7). For a material with a higher solids content, per unit volume, a greater density will be achieved.

4.4. ATTAINING LINER-LIKE PERMEABILITIES

A field program, designed to demonstrate the viability of applying the principles of sound engineering achieving liner-like cap permeabilities, was initiated at facility II. Ash from this facility, not equipped with an APC system, represented an opportunity to demonstrate the cost-effective methodology of in-situ addition of PC and of lime (CaO) to nonreactive MWC ash. The facility II ash used in the field program was subject to proprietary mechanical processing prior to placement.

4.4.1. Test Patch Description

Three test patches (TP's) were formed according to the following dimensions: $20\,ft \times 20\,ft$ (plan area) and 12 inches deep. Chemical additives and water of solubilization were added to a 7 inch depth. In-situ permeameters (of the double-walled type) were installed to a 6 inch depth, providing a

Table 4.4 Test Patches—Additive Dosages (6)

Test Patch	Additive Dosage (wt%)
No. 1	Portland Cement (PC)—10.0
No. 2	Portland Cement (PC)—6.6
No. 3	Lime (CaO)—6.6

margin depth for lateral and vertical leaching. Table 4.4 denotes respective test patch designation and additive dosage. Ashes used in the laboratory program were composite sampled from facility I, which is equipped with an APC system contributing unreacted CaO to CA (I).

4.4.2. Permeability Testing

Three different permeability tests were employed to provide sufficient bracketing of in-situ and laboratory results in support of empirical high-quality leachate results, demonstrated at the facility I ash monofill (8). In-situ results (28 days field cured) were obtained by employing a modified double-wall permeameter (based on ASTM 3385). Cores, extracted at 28 days field curing from each test patch, were subject to both triaxial permeability determinations and falling head permeability testing (9). In addition, field mixtures (of molded material from each test patch) and laboratory mixes (of each test patch additive dosage) were subject to falling head permeability testing. The field and lab mixes are reported at 14 days curing (at room temperature and humidity). The laboratory program conducted falling head permeability tests of composite samples of ashes from facility I. Permeability results of the field and laboratory programs are reported in Table 4.5.

4.4.3. Lime Additive (Concrete-like) Reactions

BA should not contain free available CaO, since it is collected upstream of the APC system. CA (I) equals BA plus the separately collected, lime-laden FA (Fly Ahs) and SR (Scrubber Residue). Since CA (II) does not reflect lime contribution from an APC system, the BA (I) should represent a similar composition. Table 4.5 reports that CA (II) + 0% exhibited a laboratory permeability of 1.9×10^{-5} cm/s, practically equivalent to raw BA (I) permeability of 1.8×10^{-5} s. Thus, the 6.6% CaO test patch CA (II) permeability results can be compared to lab program results for CA (I) with and without CaO addition. These latter ashes, obtained from facility I, contain significant inherent CaO contributed by the APC system's high stoichiometry.

Table 4.5 Permeability Comparison—Field and Laboratory Programs (6)
Field Program Permeability Results

Dosage	In-situ	Laboratory Test (Test Patch Ash)
CA(II)	–	1.9×10^{-5}
TP#1 CA(II) + 10% PC	Max 2.8×10^{-9}	7.5×10^{-7} to Max 1.13×10^{-8}
TP#2 CA(II) + 6.6% PC	Max 3.53×10^{-7}	4.15×10^{-7} to 2.0×10^{-8}
TP#3 CA(II) + 6.6% CAO	Max 6.4×10^{-6}	2.66×10^{-7} to 2.3×10^{-8}

Laboratory Program Permeability Results		
Ash	% Additive	Permeability
BA (I)	0%	1.8×10^{-5}
BA (I)	6% PC	1.5×10^{-7}
BA (I)	9% PC	1.7×10^{-8}
CA (I)	0%	5.5×10^{-6}
CA (I)	3% CaO	8.1×10^{-7}
CA (I)	6% CaO	4.2×10^{-6}

Note: All cores were tested at 28 days curing; test patch mixes and laboratory program testing were conducted at 14 days curing. Results reported in cm/s. Max = estimated maximum value.

The test patch program (Table 4.5) reports permeability results from 6.4×10^{-6} to 2.3×10^{-8} cm/s. Ignoring the lowest result as an anomaly, the CA (II) with CaO exhibits 10^{-6} cm/s order of magnitude permeability. This order of magnitude permeability reduction in the presence of free CaO suggests concrete-like behavior. The raw CA (I) permeability of 5.5×10^{-6} cm/s could reflect the presence of excess lime contributed from the operating APC system. Upon the addition of lime to CA (I) permeabilities ranging from 4.2×10^{-6} to 8.1×10^{-7} cm/s were achieved. These permeabilities agree with field measurements. Both sets of results demonstrate an order of magnitude reduction of permeability, suggesting the presence of a lime-based pozzolanic reaction.

4.4.4. Effect of Adding PC

The variation between in-situ and laboratory permeabilities reflects typical field and laboratory testing differences (10). Adding 10% PC to test patch CA (II) reduced the permeability by two to four orders of magnitude. The maximum in-situ permeability, measured within test patch No. 1, of 2.8×10^{-9} cm/s, achieved with 10% PC added to CA (II), compares similarly to a permeability of 1.7×10^{-8} cm/s, for BA (I) with 9% PC added. The maximum in-situ permeability, measured within test patch No. 2, of 3.53×10^{-7} cm/s agrees with the laboratory permeability of 1.5×10^{-7} cm/s, for BA (I) with 6% PC. Thus, the addition of 6–10% PC added to nonreactive MWC ash attained permeabilities varying from slightly greater to at least an order of magnitude less than the liner requirement of 1×10^{-7} cm/s. PC-treated, MWC ash exhibited lower permeabilities than those with CaO addition.

4.5. HEAVY METAL REDUCTION DUE TO POZZOLANIC ENCAPSULATION

As discussed in Chapter 3, the presence of lime induces pozzolanic behavior as a benefit to reduce the ashes' leachate potential. The following section discusses the implication of the field leachate and raw pH data.

4.5.1. Significance of Low pH in Field Leachate

Table 4.6 depicts that the ashes lie within the pozzolanic requirements and that the ratio of pozzolanic reactants lies within the range of analogous clean coal technology residues ((12) and Figure 4.3).

As shown in Table 4.6, the raw pH ranged from 10.91 to 11.85, while the leachate pH ranged from 6.5 to 7.4. The significant pH drop from raw ash levels to leachate values could occur due to the inherent pozzolanic behavior of lime-based MWC ash. The reduction of soluble alkalinity in ashes from resource recovery facilities suggests its consumption with pozzolanic reactants. As CaO reacts with Al_2O_3, Fe_2O_3, and SiO_2 to form pozzolanic end-products, it is unavailable as a soluble component and undetected in the leachate.

The significant pH drop from raw ash levels to leachate values could occur due to the inherent pozzolanic behavior of lime-based MWC ash. As CaO enters the pozzolanic reaction it is no longer available as a soluble component and not present in the leachate. The alkaline pH of the raw (lab-tested) ash and its similar reduction of leachate pH also may be explained by considering the pozzolanic chemistry. CaO reacts with Al_2O_3, Fe_2O_3, and SiO_2 to form pozzolanic end-products (13). MWC ash inherently reflects an alkaline pH; ASTM Water Leach testing of Hennepin Energy Resource Company's bottom ash showed pH = 9.2–9.3 (14). The reduction of soluble alkalinity in ashes from resource recovery facilities could be due to the consumption of pozzolanic reactants.

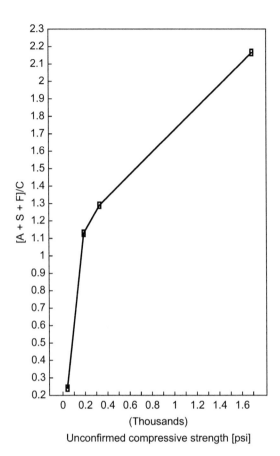

Figure 4.3 *Mineralogy vs UCS.*

Table 4.6 Pozzolanic Reactants—Leachate pH (11)

	With Lime APC
% Al_2O_3 [A]	7.39–10.30
% Fe_2O_3 [F]	3.90–18.15
% SiO_2 [S]	19.00–43.80
% CaO [C]	15.10–25.70
% A + S + F	32.76–66.85
[A + S + F]/C	1.27–3.63
pH—Ash (raw)	10.91–11.85
pH—Field leach	6.50–7.40

A study of the morphology of heavy metal-laden wastes, treated with lime, confirms their microencapsulation within the cementitious matrix (15). The author has contended that such ash should be deemed pozzolanic and recognized by regulatory authorities (16). Chapter 3, Section 3.3.2 discussed an implicit evaluation of heavy metal leachate reduction when excess lime was contributed from an operating dry scrubber. The author's interpretation of USEPA data (peer reviewed 1992) shows approximately neutral field pH's suggesting that the excess lime has reacted with the other pozzolanic constituents and is not available, i.e. not soluble (17).

4.6. FIELD MONITORING

Prior to 1990, most regulatory agencies developed their MWC characterization and management policies based on lab data. This section discusses field leachate data developed during the last four years, which should mitigate state regulatory concerns.

4.6.1. Monitoring Well Runoff Leachate (1988/1990)

Table 4.7 reports on an 18-month field monitoring study (reported by Forrester) of leachate and runoff at an MWC ash monofill (18). These averaged results, not only indicate parity to US EPA Primary Drinking Water Standards (DWS), but indicate one to two orders of magnitude lower Cd and Pb than reported by the EP Toxicity tests (19).

Such discrepancy between field and lab data questions the ability of regulatory lab leachate tests (e.g. EP Toxicity test) to realistically predict the concrete-like behavior of MWC ash. Furthermore, comparing the leachate/runoff pH of 6.7 to CA's inherent pH of 12–13 suggests a "setup" reaction. The resultant monolith precludes surface solubilization of chemical specie. Based upon the operating results presented, the MWC ash from resource recovery systems, equipped with flue-gas cleaning, when properly managed in an engineering fashion, will achieve liner-like low permeable characteristics and leachate/runoff approximating primary DWS.

Table 4.7 Leachate Runoff Collection Results

Parameter	Concentration (mg/l)	Primary DWS (mg/l)
Cadmium (Cd)	0.022	0.010
Lead (Pb)	0.007	0.050
		0.015 (Enacted Dec., 1992)
pH	6.7	6–9

4.6.2. Seven Independent Ash Monofill Field Leachate Results (1990/1991)

Recent field studies of MWC ash monofills strongly support the benign characteristics of this ash. Table 4.8, depicting independent, regulatory accepted leachate results from seven ash monofills (20,21), demonstrates that the issues raised by the USEPA-devised lab leachate tests do not occur in the real world and that actual field results prove the ash exhibits environmentally benign characteristics. Ironically, federal and state regulators are ignoring their own data. Although agreeing with my contention of incinerator ashes self-hardening, strengthening behavior (22,23,24), some agencies disregard the nontoxic results from their monitoring of local ash monofills (21).

Table 4.8 shows that in general empirical leachate values from several independent MWC ash monofills approximate Primary Drinking Water Standards. Such results should mitigate the public's concern regarding the environmental effects of exposing MWC ash during either disposal or reuse applications.

4.6.3. Long-term Effect of Ash Leachate

The leachates from the seven ash monofills offer empirical results of the benign characteristics of MWC residues. Such results, however, do not convince critics who ponder the effect of ash leachate over several years. Table 4.9 shows that the RCRA heavy metals species diminish with time,

Table 4.8 Range of Heavy Metals in Field Leachate

	EPA/CORRE[a]	Site P (1/90–9/90)	Site N (1/91–3/91)	Primary Drinking Water Standard
As	ND–0.400	<0.0020–<0.0030	<0.004	0.05
Ba	ND–9.220	1.38–2.020	2.37	1.0
Cd	ND–0.004	<0.0020–0.0033	<0.005	0.001
Cr	ND–0.032	<0.0200–0.0320	<0.010	0.05
Cu	ND–0.012	0.0114–0.212	<0.025	None
Pb	ND–0.054	0.210–0.0300	<0.008	0.015
Hg	ND	<0.0020	<0.0005	0.002
Se	ND–0.340[b]	ND	<0.008	0.01
Zn	0.0052–0.370	0.066–1.8700	<0.020	None

ND, not detected; None, no USEPA Primary Drinking Water Standards established for this specie. All concentrations in mg/l.
[a]Ref. (11); CORRE, Coalition on Resource Recovery and the Environment.
[b]Plant ZA—Se not detected in raw ash (i.e. mg/kg) or in EP TOX or TCLP. Since Se not detected in source, this result should be considered an anomaly.

Table 4.9 Long-Term Field Leachates MWC Residue

RCRA Heavy Metal	Woodburn Ash Monofill Field Leachates (mg/l) (25)				Primary Drinking Water MCLs (mg/l)
	1988	1989	1990	Feb., 1991[a]	
As	0.13–0.40	0.047–0.059	<0.01	ND	0.05
Ba	NA	ND	0.7–1.15	0.98	1.0
Cd	ND–0.0017	0.0013–0.0014	ND	ND	0.001
Cr	0.008–0.032	ND	ND	ND	0.05
Pb	0.011–0.054	0.008–0.018	<0.01	ND	0.05/0.015[b]
Hg	ND	ND	ND	ND	0.002
Se	0.12–0.34	0.24–0.33	<0.01	ND	0.01

Notes: MCL, maximum contaminant level; NA, not analyzed; ND, not detected.
[a]Per cited report, last sample representing leachate from aged ash monofill (p 22).
[b]0.015 mg/l will be enforceable Dec. 1992; present MCL = 0.05 mg/l.

supporting the environmental compatibility of ash utilization and confirming the chemical reaction of ashes' pozzolanic constituents.

These field leachates, from an aging MWC ash monofill, indicate their reduction with time. Such results show the long-term effect of exposure of the ash and its resultant potential hydrogeological impact. With the exception of barium, after four years of exposure the heavy metals' leachates were not detectable, i.e. zero for practical purposes. Significant leachate reductions appear after two to three years of aging, suggesting a slow but consistent chemical reaction. Such a reaction could be due to the MWC residues' inherent pozzolanic behavior.

Woodburn's CA was placed wet and nominally compacted with a bulldozer. Although water and compaction were applied, the optimal principles of water of solubilization (as a function of ash surface area) and optimal compaction may not have been achieved. The in-situ reaction of the free available lime with ashes' other cementitious components, therefore, occurred relatively slower. To accelerate the in-situ pozzolanic or heavy metal encapsulation behavior, the rigorous application of engineering principles should be employed.

4.7. INCREASED STRENGTH AND REDUCED HEAVY METAL LAB LEACHATE

Other investigators have stated that no relationship exists between increased geotechnical properties and reduced heavy metal leachate (26). The author disagrees with this statement and offers this and the next section to refute this contention.

Table 4.10 Nominal vs Rigorous Application Reduces Lead Lab Leachate

	Combined Ash Nominal Application (mg/l)	Combined Ash Rigorous Application (mg/l)
Pb	2.04	1.48

CA from an operating waste-to-energy facility was prepared according to the principles of compacting at optimum water of solubilization (3). Geotechnical testing involved determining the (1) optimum water of solubilization, (2) compaction (Harvard Miniature), (3) curing, and (4) structural integrity testing (13).

Setup monitoring tests (13) were conducted to determine "optimum water of solubilization". Based on these results, bottom ash and fly ash (from intermittently operating dry lime scrubber) mixture with 35% water was prepared for subsequent testing. Using the Harvard Miniature (9) apparatus, compacted cylinders were formed at a 35% optimum solubilization water content. Cylinder curing occurred over a seven-day period from August 8, 1988 to August 15, 1988. Curing during this period indicated an increase of unconfined compressive strength (UCS) with time. The 35% water sample exhibited a 0.5 ton/square foot (TSF) UCS after 24 h and a 4.5 TSF UCS after 7 days curing. Structural integrity testing was performed in conformity to the USEPA's Appendix II (27). The resultant material was taken to a New Jersey approved lab for subsequent chemical analysis.

Current USEPA federal program's results do not reflect forming test specimens with adequate water of solubilization. "This decrease (of UCS) may be attributed to increased dryness of the treated material as available moisture was depleted during the setting reaction. The addition of larger proportions of water in the S/S (solidification/stabilization) process should be investigated to provide improved strength formation" (28).

4.7.1. EP Toxicity Reduction and Increased Strength

The resultant ash, after structural integrity testing, was submitted for EP Toxicity testing. Table 4.10 compares the lead (Pb) results from split samples, reflecting nominal and rigorous application.

These results indicate that the pozzolanic behavior achieved a heavy metal EP Toxicity reduction of 27% for Pb when UCS increased from 0.5 to 4.5 TSF over 7 days of curing. Further reductions and higher strengths were likely since curing time of concrete-like material typically is tested at 28 days. By first attaining improved engineering physical properties the environmental concerns (e.g. heavy metal leachate) are ameliorated.

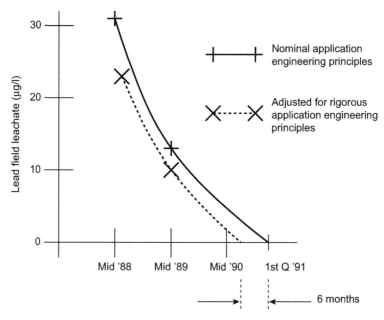

Figure 4.4 *Rigorous Application of Engineering Principles Reduces Leachate Level and Curing Time.*

4.7.2. Mathematical Predictive Relationship Derived from Woodburn Leachates

In general, Woodburn's mean RCRA heavy metal leachates can be depicted by a declining exponential curve, as expressed by the following Eqn (4.3):

$$\text{Constant} = [\text{HM}]^{-1/T} \tag{4.3}$$

where T = curing time; HM = RCRA heavy metal field leachate (ug/l) present at time, T; and Constant = slope of semilog plot and varies per heavy metal specie, ug/l = microgram per liter.

Figure 4.4 depicts mean Pb concentrations for the Woodburn actual results and for Pb levels considering the conservative 27% reduction. The higher leachate values reflect nominal application of engineering principles while the lower curve represents the observed concentrations adjusted due to rigorous application. Figure 4.4 suggests that engineering adherence should reach nondetectable levels about 6 months sooner than observed. Rigorous application could have reduced the time to reach nondetectability by 18% of the 33-month period.

Based on a semilog curve-fit of Eqn (4.3), using Pb mean leachate values from Figure 4.4, and setting the same level of Pb reduction for either case, the following equation compares curing time under the two modes:

$$T_{rig.} = (\log 3.6/\log 5.0) \times T_{nom.} = 0.8 \, T_{nom.} \quad (4.4)$$

where $T_{nom.} =$ curing time for nominal application and $T_{rig.} =$ curing time for rigorous application.

Equation (4.4) agrees with Figure 4.4 and confirms that a reduction of about 20% is expected when the engineering principles are rigorously applied. This expected reduction is conservative, since adjustment was based upon fresh CA cured for only seven days and sporadic free lime contributed from the intermittently operating APC system.

Although a nominal application of the engineering principles achieved a practical zero level of field leachate within a few years and all the ash monofill field leachates approximate Primary Drinking Water Quality (in terms of RCRA heavy metals), employing these concepts would have reduced the measurable levels while accelerating the practical zero limit of detectability. Adherence to these principles provides an additional engineering margin, which enhances ash disposal practice. These principles must be applied when the ash is substituted for traditional raw material, i.e. beneficial reuse.

Due to the recent policy of USEPA, designating MWC residues as non-hazardous, the states should consider the application of the above principles and the subsequent field results, which demonstrate the environmental compatibility of MWC residues in a landfill setting. Recognition by state agencies would facilitate beneficial reuse of MWC ashes.

REFERENCES

1. Goodwin, R.W.; "Land Disposal: Fly Ash and Sludge"; presented at American Power Conference, Chicago, Illinois, April, 1980.
2. Goodwin, R.W. et al; "Properties of Stabilized FGD Sludge"; presented at APCA Conf.; 6/25–30/78, Houston, TX.
3. Goodwin, R.W.; "Engineering and Design Aspects: Disposal and By-Product Utilization of FGD Residues"; presented at the Tenth EPA/EPRI Symposium on Flue Gas Desulfurization; Atlanta, GA; November 18–21, 1986.
4. Goodwin, R.W.; "Ash from Refuse Incineration Systems: Testing, Disposal, and Utilization Issues"; presented at MASS-APCA Conference; Atlantic City, NJ, November 3–6, 1987.
5. Jumikis, A.R.; Soil Mechanics; Robert R. Krieger Publishing Co.; Malabar, FL; 1984; pp 82–84.

6. Forrester, K.E.; Goodwin, R.W.; "MSW-Ash Field Study: Achieving Optimal Disposal Characteristics"; Journal of Environmental Engineering; American Society of Civil Engineers [ASCE], Vol. 116, No. 5, Proc. Paper 25094, Sep/Oct 1990, pp 880–889.

7. Cedergren, H.R.; Seepage. Drainage, and Flow Nets; John Wiley & Sons, NY; 2nd Edition, 1977; p 44.

8. Forrester, K.E.; "Comparison of MSW to MSW Ash Leachates Using a Risk Algorithm"; NSWMA Conference Proceedings, 1988; Boston, MA Conference on MSW, Waste Tech '88.

9. U.S. Army Corps of Engineers; Engineering Manual-Laboratory Soil Testing; EM 1110–2-1906, 1970; Washington, D.C.

10. Zimmie, T.F.; Riggs, C.O. (editors); ASTM Publ. No. 746; 1982; Permeability and Groundwater Transport; pp 56–58.

11. NUS Corp.; Characterization of Municipal Waste Combustion Ash. Ash Extracts, and Leachates; USEPA/CORRE Contract No. 68–01–7310; February 1990.

12. Goodwin, R.W.; "Meeting Clean Air Act Also Means Managing Solid Waste"; Power; Vol. 134, No. 8; August 1990; pp 55–58.

13. Goodwin, R.W.; Schuetzenduebel, W.G.; "Residues from Mass Burn Systems: Testing, Disposal and Utilization Issues"; Proceedings of the NYS Legislative Commission's Solid Waste Management and Materials Policy Conference; NYC Hilton Hotel; February 11–14, 1987.

14. Schuetzenduebel, W.; Personal Communication; Blount Energy Resource Corp.; 4/24/90.

15. Newman, A.; "Is Cement All It's Cracked Up to Be?"; Environmental Science & Technology; Vol. 26; No. 1; 1992; pp 42–43.

16. Goodwin, R.W.; "Residues from Waste-to-Energy Systems"; comments submitted to USEPA Pursuant to Proposed Amendment Subtitle C of RCRA (40 CFR Parts 261, 271 and 302); 7/31/86.

17. Goodwin, R.W.; "Resolving Environmental Concerns of High Volume Combustion Residue Beneficial Reuse – An Engineering Challenge"; Utilization of Waste Materials in Civil Engineering Construction; Inyang, H.I.; Bergeson, K.L. (editors); American Society of Civil Engineers; September 1992.

18. Forrester, K.E.; "State-of-the-Art in Thermal Recycling Facility Ash Residue Handling, Reuse, Landfill Design and Management"; presented MSW Technology Conference; San Diego, CA; 1/30–2/1, 1989.

19. Goodwin, R.W.; "Utilizing MSW Ashes as Monofill Liner"; Proceedings 1989 National Solid Wastes Forum on Integrated Waste Management; Association of State and Territorial Solid Waste Management Officials; Lake Buena Vista, FL; July 17–19, 1989.

20. Characterization of Municipal Waste Combustion Ash, Ash Extracts, and Leachates; CORRE/USEPA Contract No. 68–01–7310; February 1990.

21. Goodwin, R.W.; "Incinerator Ash – The Tip of the Iceberg"; presented at 8th Annual Conference on Solid Waste Management and Materials Policy; January 28–31, 1992; New York City; sponsored by New York State Legislative Commission on Solid Waste Management.

22. California Dept. of Health Services, 1990; "Classification of Stanislaus Waste Energy Company Facility Ash"; 2/8/90.

23. NJ Dept., Environmental Protection; "New Jersey's Ash Management Program"; presented Institute for International Research 3rd Annual MSW Incineration Ash Conference; Elizabeth, NJ; September 11–12, 1989.

24. Wiles, C.C.; "The USEPA Program for Evaluation of Treatment and Utilization Technologies for Municipal Waste Combustion Residues"; Proceedings: Municipal Waste Combustion; Air & Waste Mgt. Assoc.; Tampa, FL April 1991.

25. AWD Technologies Inc.; <u>MWC – Ash & Leaching Characterization Monofill – 4th year Study</u>; March 1992 (GR-WTE-0450).
26. Wiles, C.C. et al; "The USEPA Program for Evaluation of Treatment and Utilization Technologies for MWC Residues"; <u>Proceedings WASCON '91 Conference; Maastricht, The Netherlands</u>; November 10–14, 1991.
27. USEPA; <u>SW 846 – Test Methods for Evaluating Solid Waste: Physical/Chemical Methods</u>; 3rd Edition; September 1986.
28. Holmes, T.L. et al; "A Comparison of Five S/S Processes for Treatment of MWC Residues – Physical Testing"; <u>Proceedings WASCON '91 Conference</u>; Maastricht, The Netherlands; November 10–14, 1991.

CHAPTER 5

Operating Disposal Project Results

Contents

5.1. INTRODUCTION

Using combustion residue is preferable to its disposal. Beneficial use saves disposal costs, may yield a revenue stream, and reduces construction project costs. Implementing ash beneficial use faces obstacles. Only about one-third of CCR (Coal Combustion Residuals) is used due to outdated public work specifications and cyclical construction demands. Using incinerator ash is inhibited by the 1994 U.S. Supreme Court decision that municipal waste

Combustion Ash Residue Management
http://dx.doi.org/10.1016/B978-0-12-420038-8.00005-7

combustion (MWC) residue to be regulated as a hazardous waste (i.e. tested per Subtitle C of the Resource Conservation and Recovery Act or RCRA). This legally, but not scientifically based decision, labels incinerator ash as hazardous—creating a regulatory predisposition not inclined to dispel this misconception not to support incinerator ash utilization modes.

These MWC residue, typically suitable for a variety of construction applications, poses perceived risks due to their heavy metal content. Combustion residue beneficial use applications must answer such questions as "What happens in the long term when this ash is used? How do you know it won't cause a problem or do damage in the future?".

Monitoring of beneficial use applications are typically conducted as part of a demonstration program—addressing groundwater and fugitive emission concerns. Such demonstration programs, usually conducted for a few years, incorporate ash as a component of the final construction product material. Using an MWC residue monofill as a surrogate to evaluate the long-term, environmental effects provides a worst case (i.e. 100% ash) for end-product utilizations. An MWC residue monofill study, conducted from 1997 to 2001, addressed the long-term environmental issues of ash placement.

5.2. SITE DESCRIPTION

A 35-acre lined, MWC residue monofill (located in the Mid-Atlantic region of the USA) has been operating since 1989. This facility receives MWC residue of about 16,000 tons/month. The landfill is monitored via a series of background and down-gradient monitoring wells and raw in-situ leachate is collected from manholes. A municipal agency operates this landfill and performs all sample collection and laboratory analysis. Quarterly results are submitted to state regulators.

5.2.1. Monitoring Well Results

Background and down-gradient monitoring well results for the period 1997–2001, enable an evaluation based on the possible impact of landfill leachate to groundwater quality. Groundwater quality impact is based on down-gradient results exceeding background and on results exceeding standards. Chromium, mercury, selenium, and zinc exceeded both background levels and groundwater standards. These results suggest the landfill leachate and/or uncontrolled runoff could be adversely affecting groundwater quality.

5.2.2. Manhole Leachate Results

Manhole leachate results reflect the percolate of the landfilled ash. Chromium, lead, selenium, and zinc exceed groundwater standards and comparison to seven other ash monofills (1).

These results represent discrete, grab samples collected from manholes placed within the landfilled residue. The in-situ and down-gradient exceedances of the same heavy metals suggest a correlation between the landfill and monitoring well results.

5.2.3. Toxicity Characteristic Leachate Procedure Results

Lead (Pb) and cadmium (Cd) have frequently exceeded allowable limitations (2). Lead and cadmium toxicity characteristic leachate procedure (TCLP) results of the landfilled MWC residue show conformity to regulatory limitations; thus, they can be labeled as nonhazardous and environmentally compatible.

The typical TCLP results show agreement with regulatory maximum allowable limits for chromium, mercury, and selenium. Again laboratory results show that the MWC residue classifies as nonhazardous while exhibiting environmental compatibility.

These lab tests show conformity to regulatory criteria for chromium, lead, and selenium—suggesting environmental compatibility. Prior to landfill, the residue reflected regulatory acceptability and environmental compatibility. Since the landfilled residue exceeded groundwater and comparative levels, apparently the MWC residue quality deteriorated when landfilled.

5.2.4. Liner Integrity

The state regulatory agency requires monthly submissions of leachate volume (calculated and measured). Immediate notifications are required if the action leakage rate (ALR) exceeds 20 gallons/acre/day (based on 30-day average). Based on reported leachate volume from 1997 to 2001, the ALR was not exceeded and the liner system was deemed intact.

Although the state regulatory review documented liner integrity, groundwater levels were breached. The excesses cannot be attributed to suspect incoming MWC residue (regulatory tests showed acceptability) and the liner remained intact (permitting *de minimus* leachate). The excessive heavy metal levels must be due to an operational upset; i.e. a release of heavy metals overflowing to the groundwater.

5.2.5. Long-term Trends RCRA Heavy Metals Leachate

The discrete, in–situ leachate results do not reflect homogeneity nor continuity. In–situ leachate results can be converted into long–term trends by applying a logarithmic trend–line, is justified based on a mathematical predictive relationship derived from another operating ash monofill (3).

$$Constant = [HM]^{-1/T}$$

where T = curing time, HM = RCRA heavy metal field leachate (mg/l) present at time, T; and Constant = slope of semilog plot and varies per heavy metal species (mg/l).

For example, as shown by Figure 5.1, zinc in–situ leachate's logarithmic trend–line exceed groundwater standard over the sampling period. This long–term methodology was applied to the heavy metal in–situ leachate results.

Table 5.1 shows the long–term trends of all RCRA heavy metals and their comparison to groundwater standards. Long–term trends for mercury and selenium conform to groundwater standards, although their down–gradient monitoring well levels were excessive.

Long–term trends of cadmium, lead, and zinc exceed these standards. Long–term trends for chromium and selenium, exhibiting excessive down–gradient and in–situ results, conformed to groundwater standards.

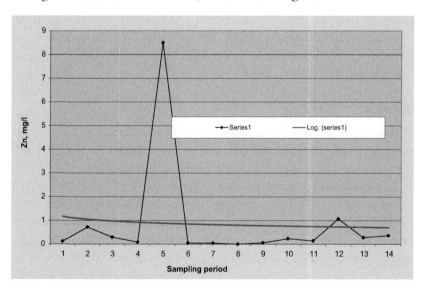

Figure 5.1 *Zinc Leachate vs Time.* (For color version of this figure, the reader is referred to the online version of this book.)

Their trend conformity supports the contention that their excesses were incident related and not systemic to the ash characteristics. Although long-term trends for cadmium and lead exceeded groundwater standards, empirical monitoring well down-gradient results did not show such expected elevated levels. The integrity of the liner, allowing minimal leachate, higher trends of cadmium and lead to deteriorate groundwater quality.

Zinc represents the only heavy metal whose down-gradient and long-term trend exceed regulatory criteria. The minimum allowable leachate, associated with the liner, could not mitigate the excessive levels of zinc.

A release occurred of regulatory acceptable ash deposited in a lined landfill.

5.3. UPSET INCIDENT

In 1994, per operator, lime was added to the residue to immobilize lead and cadmium. This lime/ash material prematurely reacted to form a concrete-like pozzolanic material that reduced leachate collector pipe opening—causing a back-up and overflow, as depicted by Figure 5.2. This operational problem have occurred due to failure to appreciate in-situ pozzolanic reaction of lime-laden ash—causing leachate back-up and overflow.

5.3.1. Residue Management—Placement

The inherent pozzolanic-like behavior of lime-laden MWC residue had been identified in 1988—providing warning time for the prudent operator to avoid premature set-up (4). The application of sound engineering

Table 5.1 Long-Term RCRA Heavy Metals Leachate Prediction

RCRA Heavy Metals Leachate	Long-Term Prediction	Groundwater Standards
Arsenic (As)	0.008	0.025
Barium (Ba)	0.008	–
Cadmium (Cd)	0.075	0.01
Chromium (Cr)	0.05	0.05
Copper (Cu)	0.65	–
Lead (Pb)	0.045	0.025
Mercury (Hg)	~0.0000	0.002
Selenium (Se)	0.002	0.01
Zinc (Zn)	0.7	0.3

Note: Concentrations expressed in mg/l.

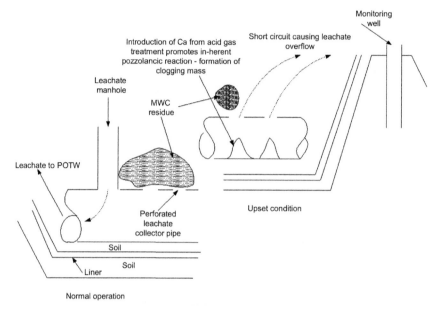

Figure 5.2 *Upset Clogging Condition.*

practice to combustion residue placement was recognized and publicized before 1994 (5,6). Avoiding the condition of premature pozzolanic reaction has been recognized occurring with coal combustion by-products (7) and analogized to MWC residues later on (8).

Sound engineering practice of combustion residue placement recognizes the relationship between achieving desired geotechnical property and optimal moisture content while maintaining adequate "water of solubilization" (or % final solids) to ensure reacting pozzolanic constituents with free available lime. Adhering to these principles requires proper placement control management—field compaction and water addition. Improved field placement (i.e. compaction, addition of dust suppression water) of residue could increase density and reduce permeability to decrease leachate rate through the buried residue (9). These practices have been embraced by Environment Canada (10).

These principles of sound engineering management of MWC residues were well-known and widely publicized. If the landfill operator had applied these principles the upset incident could have been avoided.

The MWC residue inherently was environmentally compatible and the liner was intact (allowing minimal leachate) but groundwater standards were exceeded. The release of heavy metals was due to the operator's failure to appreciate the pozzolanic nature of the ash—failure to apply well publicized and regulatory recommended engineering placement methodology.

USA regulatory officials should consider incorporating these principles into residue management recommendations, following Environment Canada's example. Recognition and implementation of these principles would confirm that incinerator ash can be properly managed—to alleviate concerns—justifying their beneficial reuse (1).

5.4. TENNESSEE VALLEY AUTHORITY COAL COMBUSTION RESIDUE DIKE FAILURE

The highly publicized, Tennessee Valley Authority's (TVA's) Kingston CCR dike failure represents another example of failure to implement sound engineering design and practice.

5.4.1. Background

On December 22, 2008, combustion residue dike ruptured at an 84-acre (0.34 km^2) coal waste containment area at the Tennessee Valley Authority's Kingston Fossil Plant in Roane County, Tennessee, USA and 1.1 billion gallons (4.2 million m^3) of coalfly ash slurry was released. The coal-fired power plant, located across the Clinch River from the city of Kingston, uses ponds to dewater the fly ash, a by-product of coal combustion, which is then stored in wet form in dredge cells. The slurry (a mixture of fly ash and water) traveled across the Emory River and its Swan Pond embayment, on to the opposite shore, covering up to 300 acres (1.2 km^2) of the surrounding land, damaging homes and flowing up and downstream in nearby waterways such as the Emory River and Clinch River (tributaries of the Tennessee River). It was the largest fly ash release in the history of United States.

In May 2009, the United States Environmental Protection Agency (USEPA) signed an enforceable agreement with the TVA to oversee the removal of coal ash at the TVA Kingston Fossil Fuel Plant in Roane County, Tennessee, where more than 5 million cubic yards of coal ash spilled last December. The cost of this clean-up is estimated at almost 1 billion dollars; TVA's predicted estimate ranges from $675 to $975 million. For example, a Middle Georgia landfill is accepting test loads of coal ash waste this week from this historic Tennessee spill. The Veolia landfill in the Taylor County town of Mauk is vying to receive the roughly 5.4 million cubic yards of waste from the spill.

5.4.2. Current Methods of Handling Coal Combustion Residues

5.4.2.1. Bottom Ash and Slag

Bottom ash collection systems often consist of water-filled tanks or hoppers located under the boiler. They vary in size, shape, and other design aspects. Water in the hopper quenches the ash and causes it to fracture into smaller sizes. It is also necessary to cushion the fall of material and protect the refractory walls of the vessel. Water is also utilized as an agent to wash or "sluice" ash out of the hopper for disposal.

5.4.2.2. Fly Ash and FGD Sludge—Landfill

Typically, fly ash is collected in the electrostatic precipitator or baghouse hoppers and conveyed by blowers to silos for temporary storage. From the silos, fly ash is discharged into pug mills where it is mixed with scrubber sludge for disposal in the on-site landfill. The silos can also be discharged dry into trucks for off-site use.

5.4.2.3. Fly Ash and FGD Sludge—Ponding

Pond containment dike slope stability depends on construction materials as well as subsurface soil conditions below the embankments and ponds. These ash landfills/ponds need to be designed to prevent localized slope failures that weaken the embankment or global failure that causes the entire embankment to rupture.

Slopes cannot be too steep. Typically slopes greater than two horizontal (H) to one vertical (V) are prone to erosion, slip failures, and global stability failures. Soil types, strength of the soils utilized for construction of these slopes, and groundwater conditions all impact slope stability.

At the Kingston site, TVA employed wet ponding rather than dry landfill. The TVA Kingston Fossil Plant coalfly ash slurry spill occurred, when an ash dike ruptured at an 84-acre ($0.34 km^2$) solid waste containment area releasing the coalfly ash slurry. According to the January 5, 2009 *Tennessean*, a dry collection system—a method that is more labor intensive—is considered more environmentally safe for waterways and groundwater than the wet method. It also was the most expensive fix at $25 million, according to TVA. The liner installation was estimated at $5 million, but TVA noted that it would set "a precedence for all other dredge cells" and "take a long time to construct".

5.5. TVA IMPLEMENTED LEAST COST RETROFIT TO DIKE—LEADING TO ITS ULTIMATE FAILURE

The cheapest option, a new dredge cell, would cost $480,000 and was a possibility for the short term, according to TVA in 2003, but could be viewed as a lateral expansion that would require the onus of a major permit modification, the update said.

In April 2005, TVA submitted a proposal for repair that would include a series of trench drains at different levels on the dike, another drain at the base of the mound and a riprap channel. Trench drains were not mentioned in the earlier 2003 options.

"Effectiveness, constructability, economics and practical experience led TVA to focus its efforts on trench drains as the preferred fix," the April 2005 TVA report said. TVA urged quick approval of its plans so it could make repairs and resume dredging. During January 2009 Public Works Hearings, Tom Kilgore (CEO, TVA) said that TVA had chosen to implement inexpensive patches instead of more extensive repairs of the holding ponds, admitting, "Obviously, that don't look good for us".

TVA is not alone in opting for less costly and traditional retrofit to coal combustion by-products (CCBs) ponding in unlined retention basins. Wet ponding of CCBs represented typical practice in the 1950s when Kingston was constructed. Dry landfill of CCBs also reflect upsets—resulting in inexpensive legal and regulatory settlements.

For example, on October 1, 2007 the Maryland Department of the Environment signed a consent decree requiring Constellation Power Source Generation Inc. and BBSS Inc. to pay a $1 million penalty and clean-up contamination from fly ash disposal. Constellation Energy, the Murphy Firm and the Law Offices of Peter Angelos announced that the parties have reached a settlement agreement regarding a class action lawsuit filed by residents in Gambrills, MD who alleged they were damaged by the use of coal ash to reclaim a sand and gravel quarry in Anne Arundel County. The settlement is subject to approval by the Circuit Court for Baltimore City. The settlement provides for: (1) the connection of 84 households, previously supplied by private wells, to public water; (2) the establishment of two trust funds to compensate affected property owners and provide site enhancements in the neighborhood; (3) the remediation and restoration of the former quarry site; and (4) the commitment to cease future deliveries of new coal ash to the quarry. The costs and benefits of these expenditures and improvements currently are estimated to be $45 million.

5.6. MEDIA AND PUBLIC FUROR

These incidents have contributed to a public media furor regarding management of CCBs.

The *New York Times* states "1300 similar dumps across the United States—most of them unregulated and unmonitored—that contain billions more gallons of fly ash and other by-products of burning coal." Further regulation is advocated. "The lack of uniform regulation stems from the EPA's inaction on the issue, which it has been studying for 28 years. In 2000, the agency came close to designating coal ash a hazardous waste, but backpedaled in the face of an industry campaign that argued that tighter controls would cost it $5 billion a year. (In 2007, the Department of Energy estimated that it would cost $11 billion a year.) At the time, the EPA said it would issue national regulations governing the disposal of coal ash as a nonhazardous waste, but it has not done so."

Similar demands are echoed by environmental advocacy groups—influencing elected officials to pressure regulatory to create more stringent requirements for CCB disposal practice.

5.7. IMPLICATIONS—MORE STRINGENT REGULATORY CONTROL

5.7.1. Federal Regulations History

In 1980, the U.S. Congress adopted the Bevill Amendment, an amendment to the RCRA. The USEPA was required to conduct a detailed study on the adverse effects of the disposal and utilization of fly ash, bottom ash, boiler slag, and other by-product materials produced from the combustion of coal.

On August 2, 1993, EPA presented its final regulatory decision on fly ash, bottom ash, boiler slag, and flue gas emission control waste, stating that these materials are not regulated as hazardous wastes under Subtitle C, and officially placed them under Subtitle D as solid wastes under the jurisdiction of individual states. EPA was to further evaluate the hazardous or toxic properties of industrial solid wastes, but at this time, CCBs were expected to remain under state regulation.

EPA preliminarily concluded in March 1999 that the remaining wastes not covered under RCRA Section C also did not warrant federal hazardous waste regulation. Environmental groups attacked this preliminary conclusion, citing public health concerns, hazardous environmental effects, and poor state regulation enforcement. A final decision was delayed several times, and at the beginning of 2000, EPA appeared ready to push for a hazardous designation. However, in April 2000, EPA determined reuse of CCBs did not warrant regulatory oversight. This landmark decision cleared the way for continued progress in the beneficial reuse of CCBs.

In May 2009, the USEPA committed to issuing proposed regulations for the management of coal combustion waste by utilities by the year's end, a senior agency official told the House Transportation and Infrastructure committee. Rules could include tightened restrictions on contaminants in wet scrubber wastewater streams. Barry Breen, acting assistant administrator for the EPA's Office of Solid Waste and Emergency Response, said that as a result of the massive coal ash spill at the Tennessee Valley Authority's Kingston plant in December 2008, the agency had embarked on a "major effort to assess the stability" of similar impoundments and other management units containing wet-handled coal combustion waste.

"This assessment has three phases: information gathering through an information request letter; site visits or independent assessments of other state or federal regulatory agency inspection reports; and final reports and appropriate follow up," he said in a testimony before the Subcommittee on Water Resources and the Environment on Thursday. "Currently, we are still in the information gathering phase and plan to begin field work in May of this year."

The purpose of the hearing was to gather more information about the disposal of coal combustion waste and water quality. USEPA Administrator Lisa Jackson promised, during her confirmation hearing, to promulgate stricter coal plant waste storage regulations capitalizing on the political opportunity of the TVA Kinston Spill will escalate with continued congressional hearings. Elected and appointed officials will find the coal industry an easy and demonized target to impose more stringent regulations.

5.8. DISCUSSION

The upset condition, causing excessive groundwater levels, could have been avoided by applying the publicized, peer-review, engineering placement principles and methodology. Operators of ash landfill facilities are urged to apply this methodology.

When implementing beneficial use applications these same principles and techniques apply. For those utilization modes requiring a soil-like consistency (e.g. structural fill, sanitary landfill cover, road construction base course), attaining optimum moisture content and maximum density achieves desired geotechnical properties. When a concrete-like, pozzolanic reaction is desired (e.g. remediation cap, stabilization/solidification of contaminated material), the optimum % water of solubilization should be introduced. Solubilizing the pozzolanic constituents capitalizes upon the ashes' mineralogy or supplements the ash with nonproprietary additive. Addition of "water of solubilization" also mitigates fugitive dusting.

USA regulators should follow Environment Canada's lead by also recommending this engineering placement methodology. Good design cannot compensate for human operator error—proper training and monitoring recommendation provided.

The incidents and consequences of TVA dike failure and spill should not indict coal-fired power plants nor the electric utility industry; unexpected costs from tens to hundreds of million dollars and public embarrassment are sufficient punishments. How to avoid such upsets should be the focus of coal-fired power plant operators and the electric utility industry.

An engineering approach, reflecting demonstrated technology and recognizing CCBs chemical and geotechnical properties, should be embraced by the electric utilities with coal-fired power plants. Commitments—to regulators—to develop and implement this approach should curb excessive requirements. Electric utilities should capitalize upon the industry-wide knowledge and submit to USEPA as regulatory approaches are being developed.

REFERENCES

1. Goodwin, R.W.; "Defending the Character of Ash"; Solid Waste & Power; Vol. 6, No. 5, September/October 1992; pp 18–27.
2. Goodwin, R.W.; "Air Pollution Control Residue Field Leachate Studies Questions Regulatory Lab Evaluation Approaches"; Proceedings of 5th International for Power Generating Industry (Power-Gen '92); Orlando Convention Center; Orlando, FL; November 17–19, 1992.
3. Goodwin, R.W.; Combustion Ash/Residue Management – An Engineering Perspective; Noyes Publications/William Andrew Publishing; Mill Road, Park Ridge, NJ, 1993; (ISBN: 0-8155-1328-3) (Library of Congress Catalog Card No.: 92-47240); pp 50–52.
4. Goodwin, R.W.; "Non-Hazardous Concrete-Like Behavior of Ash From Resource Recovery Systems Equipped with Acid Gas Treatment"; Presented at the U.S. Conference of Mayors 7th Annual Resource Recovery Conference; Washington, DC; Mayflower Hotel; March 24–25, 1988.
5. Goodwin, R.W.; Forrester, K.E.; "Engineering Management of MSW Ashes: Field Empirical Observations of Concrete-like Characteristics"; Proceedings USEPA International Conference of Municipal Waste_Combustion; Diplomat Hotel; Hollywood, FL; 4/11–14/89.
6. Goodwin, R.W.; Forrester, K.E.; "MSW-Ash Field Study: Achieving Optimal Disposal Characteristics"; Journal of the Environmental Engineering Division; American Society of Civil Engineers (ASCE), Vol. 116, No. 5, Proc. Paper 25094, September/October 1990, pp 880–889.

7. Goodwin, R.W.; "Design Optimization of a Flue Gas Desulfurization Sludge Handling System"; Proceedings of Columbia University Seminar on Pollution and Water Resources; Vol. IX 1975–1978; NJ Dept. of Environmental Protection, Bureau of Geology & Topography; Bulletin 75-C; pp I1–I18.
8. Goodwin, R.W.; "Non-Hazardous Concrete-Like Behavior of Ash From Resource Recovery Systems Equipped with Acid Gas Treatment"; Presented at the U.S. Conference of Mayors 7th Annual Resource Recovery Conference; Washington, DC; Mayflower Hotel; March 24–25; 1988.
9. Goodwin, R.W.; "Engineering Management: Ash Monofill Demonstrating Liner-Like Properties"; Presented at the Institute for International Research "3rd Annual MSW Incineration Ash Conference"; held at Sheraton Newark Hotel; Elizabeth, NJ; September 11–12, 1989.
10. Environment Canada; "Interim Recommended Practices for the Management of Solid Residue from Circulating Fluidized Bed Combustion"; Environment Protection Series Report, Quebec Canada; 1992EPS 1/PG/4.

FURTHER READING

Andracsek, R.; "The Obama EPA: A First Look"; Power Engineering; Vol. 113, No. 4, April 2009; p 10.
Dewan, S.; "Hundreds of Coal Ash Dumps Lack Regulation"; The New York Times; January 7, 2009.
Neville, A.; "Politics Surrounding Coal Ash Management"; Power; Vol. 153, No. 5, May 2009; pp 22–24.
http://en.wikipedia.org/wiki/Fly_ash.
We Energies Coal Combustion Products Utilization Handbook www.we-energies.com/environmental/ccp_handbook_ch2.pdf.

CHAPTER 6

Utilization Methodology—MWC Residues

Contents

Combustion Ash Residue Management
http://dx.doi.org/10.1016/B978-0-12-420038-8.00006-9

6.1. INTRODUCTION

This chapter is retained to reflect the technical issues surrounding utilization of municipal waste combustion (MWC) residues for application by developing nations—avoiding USA political rhetoric and technically unjustified regulation.

Empirical evidence from several investigations of ash landfills demonstrate environmental compatibility. The high quality leachate/runoff associated with MWC ash monofills implies that ashes' beneficial use should be considered and evaluated by regulators. The US environmental protection agency's (USEPA's) early policy focused on regulating MWC residue disposal rather than their beneficial use. The information contained in this chapter should assist developing nations' regulatory agencies in developing sound technical policy for MWC residue beneficial use.

The USEPA's nonhazardous regulation of MWC residues, i.e. Subtitle D, suggests that it falls into the category of solid waste. This category includes newspaper, glass, etc., traditional solid wastes, which are deemed suitable for recycle. Broadly interpreted, USEPA's classifying MWC residues as Subtitle D or solid waste implies their suitability for recycle or beneficial use.

6.2. GENERAL ECONOMICS—DISPOSAL VS BENEFICIAL USE

Land disposal of these residues increases the operating costs of the waste-to-energy projects while subjecting the installation to continued monitoring of its residue disposal site. Developing by-product utilization scenarios for these residues would significantly mitigate these annual costs and greatly reduce the amount of land disposed material.

Consider a typical waste-to-energy facility having a daily production capacity = 1500 tons/day (TPD). Assuming that dry acid gas removal is required and implemented, the incinerator residue and air pollution cleaning (APC) wastes constitute 20% of the feed rate (conservative assumption). At an operating period = 290 days/year, the incinerator residue and APC waste annual generation rate = 87,000 tons/year. At a conservative unit disposal cost of $20/ton, the annual disposal expense = $1.74 million. (Costs expressed in 1998 $.)

As previously discussed, the MWC residue should exhibit pozzolanic behavior due to its high lime content. This pozzolanic characteristic suggests the by-product utilization of MWC residues/APC wastes. Potential specific utilization modes can be identified by referring to the successful experience of waste from coal-fired power plants.

6.3. GENERAL UTILIZATION APPROACH

The author's experience with coal power plant utility wastes suggests that general physical and chemical characteristics provide the first step in determining the specific utilization modes for MWC residues. Initial investigations should determine the particle size distribution and chemical composition of residues; comparing them to application end-product specifications, e.g. American Society Testing and Materials (ASTM).

6.3.1. Specific Utilization Categories

Table 6.1 depicts the usage distribution for each type of coal ash, i.e. fly ash (FA), bottom ash (BA), and boiler slag.

6.3.2. Chemical Composition

A comparison of MWC residue to Portland cement indicates the similarity of chemical composition (Table 6.2).

Cement raw materials are primarily composed of silica, alumina, and iron oxide; these minerals are thermally reacted with lime to form Portland cement clinker. This clinker is the sintered product of burning the lime-based raw material with the aluminum-silicate-based mineral. The clinker is ground yielding the final Portland cement.

Table 6.1 Coal Ash Utilized as By-Product Type of Coal Ash

End-Product Use	Fly Ash	Bottom Ash	Boiler Slag
Mixed with raw material before forming cement clinker	7%	–	–
Mixed with cement clinker or mixed with pozzolans cement (Type I–P)	5%	–	–
Partial replacement of cement in concrete or concrete products	25%	–	–
Lightweight aggregate	2%	3%	–
Fill material: roads, construction sites	20%	20%	8%
Stabilization: roads, parking areas	3%	5%	2%
Filler in asphalt	6%	–	–
Ice control	–	22%	13%
Blast grit/roofing	–	–	48%
Masonry: mortar and grout, brick	3%	24%	22%
Uncategorized	29%	26%	7%

Table 6.2 Chemical Composition—MWC Ash vs Portland Cement

Component	Composition of Portland Cement		MWC Residue/ APC Waste (%)
	Cement (%)	Clinker (%)	
SiO_2	18–24	21.7–23.8	24
Al_2O_3	4–8	5.0–5.3	6
Fe_2O_3	1.5–4.5	0.2–2.6	3
CaO	62–67	67.7–70.8	37

6.3.3. Particle Size and Distribution

Besides chemical composition, the physical characteristics establish suitability for a particular by-product utilization application. The material's particle size and distribution establish its suitability for a specific end-product substitution.

6.3.3.1. Cement Fineness

The fineness of cement is a major factor influencing its rate of hydration since the reactions involved occur at its interface with water. Ordinary Portland cement (Types I and II) has a surface area between 3000 and $3500\,cm^2/g$, while the surface area for rapid hardening Portland cement (Type III) ranges from 4000 to $4500\,cm^2/g$. ASTM specifies a minimum surface area (as determined by the air permeability test, i.e. Blaine) of $2800\,cm^2/g$ for all types of Portland cement, except Type III for which a minimum surface area is not specified. Most cements pass a No. 200 sieve; these particles are smaller than $74\,\mu m$. The most active part of cement is the material finer than $10–15\,\mu m$. Furthermore, a fine grind of cement clinker is able to coat the surfaces of aggregates more completely than coarse material enabling intimate contact of the resultant paste or mortar. Since the reaction between cement and water occurs only at the solid particles' surface, the accumulation of large particles on the surface of unreacted material would hinder the overall reaction. The finer the cement particles and the greater its surface area, the faster is the rate of hydration and the greater is the proportion of the cement which reacts within the paste.

6.3.3.2. Aggregate Fineness

The mineral aggregate usually occupies from 70–75% of the total volume of mass of concrete. In order to provide a dense packing in a concrete mass, the aggregate must be suitably graded from fine to coarse. The smallest particles classified as coarse aggregate are defined as 0–5% retained on a No. 8 sieve or

Table 6.3 Mineral Aggregate Size Limitations

Sieve Opening	Percent Passing
9.5 mm (3/8 in)	100
4.75 mm (No. 4)	95–100
2.36 mm (No. 8)	80–100
1.18 mm (No. 16)	50–85
600 μm (No. 30)	25–60
300 μm (No. 50)	10–30
150 μm (No. 100)	2–10

material having a size greater than 0.0937 in. In general ASTM C-33 specifies the fine aggregate in accordance with the limits shown on Table 6.3.

This specification (Table 6.3) also pertains to fine aggregate for shotcrete or gunite mortar, which is the pneumatic application of cement and/or aggregate. If the aggregate is too fine, it produces a weak coating subject to excessive shrinkage. If it is too coarse, the amount of rebound (or the material which bounces back from the applied surface) will be excessive, and a rough textured surface will result. ASTM No. C-144 specifies the following gradation limitations for aggregate used in masonry mortar as shown by Table 6.4.

Table 6.4's classifications both pertain to ASTM's specifications for fine aggregate for masonry grout (C-404). C-33 applies to fine aggregate size No.1 and C-144 applies to fine aggregate size No.2.

6.4. ANALOGOUS CHEMICAL COMPARISON

Table 6.5 compares the chemical composition of MWC ash to oil shale ash. Between 4% and 8% gypsum was added to oil shale ash to compressive strength reaching 4100 psi (1). By analogous comparison, Table 6.5 suggests that approximately 15% lime should also be added to the MWC ash, assuming a dry lime scrubber. Based on this oil shale analogy, the resultant material would satisfy the specified (ASTM C-593) minimum compressive strength 600 psi for a pozzolan. Although lime addition also may be required to achieve parity with Portland cement, based upon the chemical comparison a high potential exists for utilization of MWC ash as a cementitious by-product.

Analogous comparison of MWC residue to other materials, based on a chemical composition and particle size distribution, is a valid screening tool. For example, Figure 4.3 related the significant mineralogical constituents to

Table 6.4 Fine Aggregate Size Limitations

Sieve Opening	Percent Passing
4.75 mm (No. 4)	100
2.36 mm (No. 8)	95–100
1.18 mm (No. 16)	70–100
600 μm (No. 30)	40–75
300 μm (No. 50)	10–35
150 μm (No. 100)	2–15
75 μm (No. 200)	–

Table 6.5 Chemical Comparison to Analogous Materials

		Residue/APC Waste	
Component (wt%)	Oil Shale Ash	Mass Burn	RDF
SiO_2	20	24	37
Al_2O_3	8	6	4
Fe_2O_3	4	3	5
CaO	50	37	43

Note: RDF, refuse derived fuel.

unconfined compressive strength (UCS). This curve was developed for the US Department of Energy's (DOE's) clean-coal technology (CCT) residues and based on their data. MWC residue's predicted mineralogy range, including sodium, chlorides, and loss on ignition (LOI), should be determined. Knowing the salient mineralogy and subspecies, which retard pozzolanic strengthening and heavy metal encapsulation, provides valuable screening tools in selecting combustion and pollution control technologies.

The author urges incorporation of utilization concepts during the planning and siting phases of resource recovery facilities. Besides recognition of the physical–chemical characteristics of the expected MWC residue, the following sections describe a methodology to identify high potential end-uses based upon site-specific techno-economic factors.

6.5. PRIORITIZATION OF POTENTIAL UTILIZATION MODES

This prioritization evolved from the DOE CCT residue review. It allows the combustion energy facility to evaluate the most likely by-product scenarios. This section discusses the key economic factors in determining by-product utilization viability.

Table 6.6 Preliminary Site-Specific Market Screen

Process	End-Use	Likely Region	Raw Material ($/ton)
AFBC–BA	Soil Stab.	Bituminous	2–10[a]
	Road Constr.		3–5[b]
AFBC–FA	Road base	Bituminous	2–10[a]
	Soil Constr.		3–5[b]
Dry FGD	Soil Stab.	Lignite; bituminous	2–10[a]
FA	Fill		3–5[b]
	Grout		48.85[c]
LFI–FA	Soil. Stab.	>1.5% Sulfur	2–10[a]
		retrofit	3.35[d]
	Road base		2–10[a]
	Fill		3–5[b]
Calcium injection	Soil Stab.	Possibly >1.5 sulfur;	2–10[a]
FA	Grout + mortar	Likely <1.5% Sulfur	3–5[b]
			48.85[c]

Note: Stab., stabilization; Constr., construction.
[a]Earth borrow.
[b]Aggregates.
[c]Cement.
[d]Sand and gravel.

Table 6.6 lists a prioritization of potential by-product utilization modes, based upon technical commonalities between the Electric Power Research Institute and DOE studies. This listing offers electric utilities and other combustion energy industries a starting point in their consideration of most likely utilization scenarios.

An evaluation of the referenced reports (see "Clean Coal Residue Bibliography") arrived at the menu of "common utilization modes". Technical compatibility formed the basis for these high ranking by-product utilization modes. Just technical conformity to end-user specifications insufficiently motivates a manufacturer to entertain substituting a waste for traditional raw materials. Highly favorable potential cost savings must exist before end-users seriously consider by-product utilization. The most compelling economic motivation, supporting waste utilization, is the deficiency of locally available traditional raw materials. For instance, lime and cement kiln locations plus National Lime Association's Deposit Map (which are readily available) depict the inadequacy of local limestone deposits. Using such information combustion energy facilities can easily identify paucities of potentially substitutable raw materials.

Transportation cost represents another significant economic driving force. The transport cost of the by-product must be less than the mined or process and transport cost of the traditional raw material. Combustion energy facilities should develop site-specific transport costs, which vary considerably according to vehicle type and terrain.

These modes reflect the most likely near-term utilization scenarios. Superimposing the most likely areas/regions of implementation and the raw material deficiency and/or transportation cost could further rank the most economically viable utilization mode for a given clean-coal or flue gas cleaning technology. Table 6.6 provides a preliminary guide to evaluating the proposed by-product utilization scenarios.

These modes are considered as the highest potential modes and should be addressed via a superposition of site-specific marketing factors (e.g. deficiency of raw materials, most likely implementation of advanced coal combustion and conversion processes as a function of fuel and as a function of new plant vs retrofit).

6.5.1. Expected Variation Range of Pozzolanic Reactants

Lime (CaO) in the presence of silica (SiO_2), alumina (Al_2O_3), and iron oxide (Fe_2O_3) forms sulfoalumina hydrates (ettringites), and calcium silica hydrate (tobermorite), which are the pozzolanic end-products.

Figure 4.3 demonstrated the significance of the ($Al_2O_3 + Fe_2O_3 + SiO_2$) to CaO ratio in achieving optimal geotechnical properties. Similarly technical conformity to end-user specifications requires a range of consistency and defines a mineralogy limit. Coal corings, test burns, and ultimate/proximate analyses reveal the variations in $Al_2O_3 + Fe_2O_3 + SiO_2$. The CaO range depends on the stoichiometric feed rate, which varies with sulfur dioxide (SO_2) level. Knowing the variation of these constituents facilitates predicting their compositional range in the waste by-product.

6.5.2. Expected Variation in Physical Characteristics

Particle size distribution and surface area represent the salient physical parameters to achieve end-user specification conformity. In general these factors reflected consistency for a given facility and CCT/AGC process. In some instances inconsistency occurred in particle size distribution, e.g. nonrecycled calcium spray dryer FA. Controlling size distribution and, concomitantly, surface area can be accomplished by preprocessing the residue. Depending on the expected variation and specification requirement screening and/or agglomeration could be implemented. Calcium spray dryer and LFI-FAs (Limestone Furnace Injection - Fly) may require agglomeration for use as an aggregate.

6.5.3. Degree of By-product Variation

Some end-uses reflect minimal requirements, e.g. fill. These applications require the least preprocessing, but they command a lower market (see Table 6.6). Conformity to more restrictive by-product specifications (e.g. cement/grout) could mandate incorporation of a preprocessing step. A higher market price for such material becomes justified based on the energy combustion facility's preprocessing expense and the end-user's traditional cost. Table 6.6 shows an order of magnitude greater price for cement/grout than for less restrictive raw materials.

6.5.4. Screening Menu Utilization

The following preliminary testing menu is proposed to evaluate a given CCT/FGC residue for by-product utilization:
- ultimate analysis of mineralogy including CaO, LOI, chlorides,
- geotechnical properties, UCS, coefficient of permeability,
- surface area (m^2/g), and
- percent water of solubilization.

6.5.5. Quality Control

Yielding a material for by-product utilization imposes a different set of operating conditions than for disposal. Disposal requires no preprocessing step at the power plant (other than addition of dust-suppression water). Engineering CCT/AGC residue land disposal management consists of percent water of solubilization addition and optimal compaction; such controls are neither economically nor operationally burdensome. Similarly, yielding a material suitable for less restrictive end-uses requirements imposes minimum product control demands. Conversely, conformity to restrictive end-use requirements entails greater operator attention to residue product characteristics. Closer tolerance to end-use specification demands operator vigilance. Introducing online monitors (e.g. particle size detectors) and performing laboratory analyses will facilitate maintaining specification conformity. Combustion energy facilities opting for higher market by-product utilization scenarios should recognize the quality control demands.

6.6. SPECIFIC UTILIZATION MODES

Since this material exhibits concrete-like and environmentally benign behavior, using MWC ash rather than throwing it away would save money and reduce demand on traditional raw materials. Adopting ash beneficial use avoids disposal costs while realizing unilateral economic benefit. Achieving superior end-product performance and significant cost savings constitute the elements of a successful beneficial use program.

6.6.1. MWC Residue as Raw Material Substitute—Portland Cement

MWC residue reflects a mineralogy similar to Portland cement clinker. ASTM cement product specification (ASTM C-618) requires that the total of $SiO_2 + Fe_2O_3 + Al_2O_3$ contain a minimum range of 50–70% by weight. ASTM does not provide a specification for raw material Portland cement manufacture. Table 6.7 compares the mineralogy of MWC residues to cement clinker and to conventional and advanced SO_2 conversion and/or coal combustion.

Only 7% of the conventional coal combustion ash is used for Portland cement manufacture. Such a low utilization may be attributed to a somewhat low CaO content in conventionally fired ash compared to residues from advanced SO_2 conversion and/or coal combustion systems. Dry flue gas desulfurization (FGD), LFI, and AFBC reflects clean-coal technology in terms of advanced SO_2 conversion and/or combustion. Portland cement represents their high potential utilization option (2). Given the favorable mineralogy and compatible surface area of MWC residues compared to typical raw materials and analogous ashes, up to approximately 71% substitution could be expected. Based on a typical 2000 TPD waste-to-energy facility generating 500 TPD of residue and assuming a 71% substitution, one cement plant could consume all the ash from five such plants (Table 6.7).

6.6.1.1. Impurities

Since the conceptual considerations appear encouraging, research and development efforts appear justified. Such efforts should include the possible adverse effect of soluble impurities. Table 6.8 reports constituent/impurities based on ASTM product specifications, e.g. ASTM C-595, ASTM C-593. To ensure end-user acceptance and product conformity, further testing of MWC residues according to ASTM procedures is recommended.

Table 6.7 Raw Material Substitute—Portland Cement

	Al$_2$O$_3$	CaO	Fe$_2$O$_3$	SiO$_2$	LOI	(m^2/g)
Coal-fired FA	25	1	12	54	5	0.55
Dry FGD FA	9	25	4	21	4	6.85
LFI-FA	17	38	12	16	11	4.25
AFBC-FA	15	23	19	15	13	23.9
MWC ash	6	37	3	24	<10	0.38
Cement clinker	6	62	4	22	6 (Max)	NA

Table 6.8 Potential Impurities—MWC Ash

	MWC Ash	
Constituent	With Lime APC	Limit per ASTM
A + F + S (%)	32.76–66.85	50–70
Sulfur as SO$_3$ (soluble—%)	ND–0.05	3.0–5.0
Sodium as Na$_2$O (soluble—%)	0.02–0.06	1.5
Water-soluble fraction	0.64–6.58	10.0

Table 6.9 Nonreactive MWC Residue Suitable for Road Construction

Site	% Ash	PENN DOT/Technical Performance
Philadelphia, PA (1975)	50	Acceptable
Delaware Co., PA (1975)	50	Acceptable
Harrisburg, PA (1976)	100	Excellent (fused residue)

Note: PENN DOT, Pennsylvania Department of Transportation.

In addition a practical chloride limitation of 4% by weight should be considered. Based upon extrapolation, a soluble chloride concentration of 0.0034–0.034% has been derived. Therefore, these derivative soluble impurities in MWC ash appear to satisfy ASTM allowable concentrations (Table 6.8).

6.6.2. Road Construction Applications

In the past, MWC residues have been utilized for road construction (3). Incinerator ash was tested at a few Pennsylvania road construction sites.

Conclusions derived from Table 6.9 and Ref. (3) can be summarized as: (1) LOI < 10%—eliminate organics; (2) achieve ASTM specifications;

(3) limit application to 50% ash and 50% residue; and (4) minimize fine particle component, i.e. eliminate FA.

6.6.2.1. Particle Size Restrictions

Besides chemical composition, potential end-uses require specific particle size distribution. Table 6.10A depicts the size distribution of BA and FA. A comparison shows potential uses of BA as coarse highway aggregate (ASTM D-448) and of FA as fine cement aggregate (ASTM C-33); additional segregation would be required. Such segregation could yield approximately 75% of the BA as suitable for coarse highway aggregate and 25% of the FA as suitable for fine cement aggregate. Separating (i.e. screening) coarser (>3/8–3/4 in) material from BA improves the combined ash (CA) characteristics and enhances recycle potential of the coarser residues. California bearing ratios (CBR) of nonreactive ash reached approximately 40%; i.e. suggesting that 6 in could be used in a pavement subbase (4). As showed by Table 6.10B, the FA, BA, and CA size distributions appear suitable as soil aggregate, for paving application (ASTM 1241). Such uses may not require additional size segregation.

6.6.3. BA as Cover for Sanitary Landfill

Cover for sanitary landfills eliminates exposure of the landfilled MSW. Eliminating such exposure, nuisances such as fugitive litter, vermin attraction, and erosion are prevented. Furthermore, the cover material effects the passage of precipitation through its thickness (i.e. percolation) and across its surface (i.e. runoff). The percolation through the cover material directly impacts the moisture content of the underlying MSW and, so affects the biodegradation of the buried waste. The biodegradation or decomposition contributes to the generation of methane gases and the buried waste's structural consolidation. Percolation also is related to the quantity of leachate produced.

6.6.3.1. Types of Sanitary Landfill Cover

BA is proposed as cover material for an MSW landfill. Three cover applications are considered: (1) daily cover; (2) side cell intermediate cover; and (3) interim final cover. Daily cover is applied after a working day at 6 in thicknesses. Intermediate cover is applied in 6 in lifts to a thickness of 1 ft when the working area will be inactive for 1–3 months. Interim cover pertains to the material that becomes a component of the final cover of the landfill. Interim cover is applied in 6 in lifts to a 2 ft thickness.

Table 6.10A Potential Uses of MWC Residue—Additional Segregation
Bottom Ash as Coarse Aggregate—Highway Construction

Sieve Size	Percent Finer	
	ASTM D-448	Mass Burn Bottom Ash
2 in (50 mm)	100	100
1.5 in (37.5 mm)	95–100	88
1 in (25 mm)	–	–
0.75 in (19 mm)	35–75	60
0.5 in (12.5 mm)	–	–
0.38 in (9.5 mm)	10–30	33
No. 4 (4.75 mm)	0–5	24

	Fly Ash as Fine Aggregate—Cement	
Sieve Size	Percent Finer	
	ASTM C-33	Mass Burn Fly Ash
0.38 in (9.5 mm)	100	74
No. 4 (4.75 mm)	95–100	50
No. 8 (2.36 mm)	80–100	37
No. 16 (1.18 mm)	50–85	32
No. 30 (0.59 mm)	25–65	29
No. 50 (0.297 mm)	10–30	24
No. 100 (0.149 mm)	2–10	18

Table 6.10B MWC Ash as Soil—Aggregate Subbase, Base, and Surface Courses
(Minimal Segregation)

Sieve Size	Percent Finer	
	Type I-C ASTM D-1241	Mass Burn Fly Ash
1 in (25 mm)	100	100
0.38 in (9.5 mm)	50–85	74
No. 4 (4.75 mm)	35–65	50
No. 10 (2.0 mm)	25–50	36
No. 40 (0.42 mm)	15–30	24
No. 200 (0.074 mm)	5–15	17

	Percent Finer	
Sieve Size	Type I-B ASTM D-1241	Mass Burn Combined Ash
2 in (50 mm)	100	100
1 in (25 mm)	75–85	80
0.38 in (9.5 mm)	40–75	50
No. 4 (4.75 mm)	30–60	30
No. 10 (2.0 mm)	20–45	24

6.6.3.2. Technical Requirements for Cover Material

Traditional cover materials range from using sand as daily cover, sandy clay as intermediate cover, and clay/silt as interim cover. The engineering requirements for each type of cover material may be categorized according to its permeability and/or particle size distribution. Table 6.11 summarizes such criteria.

6.6.3.3. Substitution of MWC BA

Previous work in New England using coal ash as sanitary landfill covers offers the prospect for using MWC BA in the same applications (5). Based upon prior work, BA conforms to the particle size requirements (6). Although the typical BA exhibits a permeability of about 10^{-5} cm/s, such values occurred at 95% modified Proctor compaction. By applying a standard Proctor compactive effort of 85%, the resultant permeability should be increased—approaching the daily cover requirement. The BA generated from a 2000 TPD waste-to-energy facility will satisfy the daily, intermediate, and interim cover requirements of a sanitary landfill servicing approximately 100,000 people. In the Mid-Atlantic region, final capping, composed of bentonite clay, costs $160/CY (f.o.b.) or approximately $300/CY (delivered) (7).

6.6.3.4. Daily and Interim Sanitary Landfill Cover

Daily and interim cover is applied to prevent dusting, control vermin, and provide for some passage of moisture to the buried MSW. As a general guideline, New Jersey (NJ) suggests a particle size distribution of <3 in to a maximum of 10% passing a No. 200 sieve. New York (NY) limits the percentage of fines to 5% passing a No. 200 sieve. NJ and New Hampshire (NH) recommend a maximum permeability of 10^{-3} cm/s for daily cover.

Table 6.11 Engineering Criteria—Sanitary Landfill Cover

Cover Type	Permeability	Size
Daily	10^{-3} cm/s	100% < 3 in
		Max 25% < No. 100 sieve
		Max 10% < No. 200 sieve
		(NJ)
		Max 5% < No. 200 sieve
		(NY)
Intermediate	10^{-4} to 10^{-5} cm/s	
Interim	Same as intermediate	

NH allows a lower permeability of 10^{-5} cm/s for interim cover. MWC BA conforms to such requirements.

6.6.3.5. *Effect upon Leachate and Biological Activity Rates*

McEnroe and Schroeder (8) have shown that the leakage or leachate rate through the drain layer is related to its degree and depth of saturation. With BA's lower permeability of approximately 10^{-4} to 10^{-5} cm/s compared to typical daily and intermediate cover material permeabilities reaching 10^{-3} cm/s—the amount and depth of saturation will be reduced. Therefore, the underlying head and leachate or leakage rate diminishes. Therefore, the rate of flow from underlying cells and eventually to the final liner is reduced. Using less permeable BA as daily and intermediate cover hydraulically reduces the leachate/leakage flow rate.

Moisture content of MSW promotes biological activity in terms of gas production and consolidation/settlement rates (9). Since the saturation and transfer rate of leachate would be reduced, due to the presence of less permeable BA, both gas and settlement rates should more quickly reach an equilibrium or steady-state condition. Controlling these rates mitigates safety and cracking issues.

6.6.3.6. *Effect upon Leachate Quality*

Gray (10) demonstrated the improvement of MSW leachate quality upon passage through a layer of coal/wood ash. Both organic and heavy metal contaminant reductions resulted. One-third reductions of biochemical oxygen demand and chemical oxygen demand were observed, while Cd and Pb were reduced by 81–100%. A compositional and physical analogy was developed between coal-fired ash and MWC residue (11). BA's surface area is approximately $2 \, m^2/g$. (12) and typifies granular activated carbon (GAC). GAC media remove organics via adsorption. BA's alkaline pH (>9) should mitigate the growth of deleterious microorganisms. The mechanisms of adsorption and biological inhibition could account for the expected reductions of organics and heavy metal contaminants.

6.7. BY-PRODUCT UTILIZATION CONCEPT—ECONOMICS

Establishing a scenario for by-product utilization would reduce disposal costs and offer the potential for revenue from the sales of the waste material. Demonstrating the concrete-like characteristics of MWC ash and its suitability as a self-liner, suggest applying this chapter's engineering principles

Table 6.12 Stabilization Cost Analysis of Commercial Additives

Chemical Additives	$/Ton	Comment
Portland cement	75	–
CaO (pebble lime)	60	–
Lime kiln dust	12	50% Reactive
Cement kiln dust	12	50% Reactive
Coal-fired fly ash	3	–
Gypsum	45	Purity = 87–90%

Table 6.13 Cost Comparison: By-Product Utilization vs Disposal

Basis Utilization	
Capital equipment investment	$5.5 MM
Amortization (CRF: 10%/year @ 10 year)	$0.9 MM/year
Chemical additive (dosage @ 10%)	$80/ton
Operation & maintenance	35%
Contingency	15%
Process cost = $32/ton vs disposal cost = $75–110/ton	

Note: CRF, capital recovery factor; costs in 1998$.

to utilization concepts. Obtaining by-product properties may be accomplished by seeding the MWC residue with standard additives. Table 6.12 tabulates the chemical additive unit costs used in developing a stabilization treatment cost matrix. Adding commercially available additives to MWC ash would increase operating cost by $1.60/ton of MWC ash to $4.20/ton of MWC ash. This matrix was based upon a typical waste-to-energy facility, as shown below:

- 1500 TPD capacity
- 500 TPD total ash
- BA = 85% by weight = 425 TPD
- CA = 15% by weight = 75 TPD
- 300 operating days/year.

By applying the principles of optimizing (1) particle size distribution, (2) % water of solubilization, (3) chemical additive dosage, and (4) degree of compaction or densification, a conceptual utilization system was preliminarily engineered. Table 6.13 presents a conservative budgetary estimate for a utilization plant augmented to a resource recovery facility. The unit process cost of $32/ton of ash (about $11/ton of MSW) is one-third to half the cost of ash monofill disposal in NJ. Rather than expend resources to discard

MWC residue, the waste-to-energy field (private and public sector) is urged to implement by-product concepts. By implementing the above utilization concept, savings of \$12–16/ton of MSW could be realized. Just donating the processed ash could save millions of dollars per year. (Costs are expressed in 1998\$.)

6.8. IMPLEMENTATION OF MWC ASH BENEFICIAL USE

During the siting phase of a resource recovery facility, developing a "prioritization of potential utilization modes" (Section 6.5) is recommended. Based on experience, potential end-users will accept combustion residues as substitutes for their traditional raw materials when both significant cost savings and superior end-product performance are demonstrated. Achieving enhanced engineering properties plays a predominant role in beneficial use applications. Adhering to the engineering principles ensures attaining the engineering criteria of high durability and enhanced product performance. The engineering approach realizes improved end-product quality while providing additional margins for heavy metal reduction.

6.8.1. Proposed Coproduct Evaluation Procedure

Samples to be lab evaluated should reflect that substitution portion capable of achieving superior construction end-product performance. An optimal percent substitution is first determined to reach improved material quality. Conformity to the engineering requirements establishes the optimal substitution menu. Why subject combustion residue substitution end-product to environmental testing unless it first exhibits equivalent or superior performance? Such common sense were not applied; USEPA studies subject the samples to environmental testing without having attained adequate engineering reuse properties (13). A coproduct evaluation methodology, to achieve a nonregulated reuse mode (depicted by Figure 5.1), is recommended as a slight, but significant, correction to future regulatory laboratory MWC ash evaluation programs.

Figure 6.1 represents a proposed methodology for regulators to evaluate MWC residues for utilization. The first step identifies the high potential end-uses. The second step conducts a lab program for these end-uses establishing the fraction of MWC residue, substituted for traditional raw material, yielding a superior performing end-product, i.e. the engineering first criteria. Upon determining the optimal substitution percentage, regulatory lab leachate tests can be conducted.

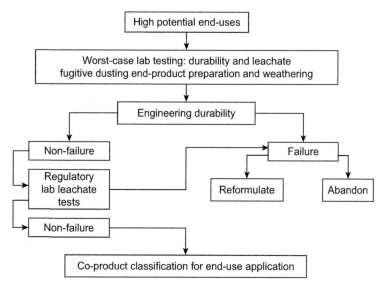

Figure 6.1 *Proposed Regulatory Methodology Coproduct Classification. By Elsevier.*

The inherent approach of this proposed methodology reflects the evaluation of end-products with residue substitution based upon initial engineering demonstration of equivalent or superior end-products prior to environmental evaluation. Environmental evaluation of end-products with substituted combustion residue should occur only when the engineering testing demonstrates enhanced structural properties, i.e. strength, permeability, and durability.

Substitution of MWC residue for traditional raw materials has the potential to be realized. Such substitution can achieve superior engineering (product quality) performance, while satisfying regulatory (environmental evaluation) concerns. An engineering analysis of durability (i.e. weathering factors) investigates potential constituent release of either consistency reuse application at respective substitution ratios.

6.8.2. Road Construction Applications—Engineering Considerations

Gradation comparison shows potential uses of BA as coarse highway aggregate (ASTM D-448) and of FA as fine cement aggregate (ASTM C-33); additional segregation would be required. Such segregation could yield approximately 75% of the BA as suitable for coarse highway aggregate and 25% of the FA as suitable for fine cement aggregate. Separating (i.e. screening) coarser (>3/8–3/4 in) material from BA improves the CA characteristics and enhances recycle potential of the coarser residues. CBRs of nonreactive ash reached approximately 40%; i.e. suggesting that 6 in could be used in a pavement subbase. The MWC ash size distributions appear suitable as soil aggregate, for paving application (ASTM 1241). Such uses may not require additional size segregation.

6.8.3. Road Construction Application—Environmental Considerations

Applying the engineering first determination for a proposed construction application implies that some fraction of ash is used to substitute for traditional raw materials. A highway, for example, would not be built of 100% ash but some lesser percentage based on achieving better than traditional raw material performance. The field results (from Chapter 4) show the environmentally benign effect of MWC residue, which could be compared to building a highway of all ash. Since some lesser percent will be used for engineering performance, the book's results represent worst-case. Under this worst-case scenario, we have shown approximation to drinking water (in terms of heavy metals) and reduction of Resource Conservation and Recovery Act or regulated heavy metals (with time), the highway built, with some fraction of MWC ash, should be considered to have less of an impact than these field results.

Tables 4.8 and 4.9 reflect the worst-case scenario of building a road of all MWC ash. The proposed protocol, however, first identifies that fraction of ash capable of achieving superior performance when compared to a control specimen of traditional raw material. The field data presented in this chapter can be represented as the "worst-case" of using 100% ash substitution in construction applications. A practical risk-assessment of beneficial use of MWC ash should incorporate these types of empirical field data as opposed to predictive models based on just lab data.

Conversion of municipal waste through resource recovery plants to produce energy and recycling of incinerator residues as construction materials will realize significant reductions of each community's solid waste expenses. By supporting recycling of municipal refuse through source separation, resource recovery, and ash reuse, our tax burden could be relieved and we could avoid an increase in local taxes. Rather than raise taxes, our elected officials and their appointees should consider the recycling of municipal solid waste into energy and the recycling of incinerator ash into construction materials.

REFERENCES

1. Bentur, A.; Grinberg, T.; "Modification of the Cementing Properties of Oil Shale Ash"; Ceramic Bulletin; Vol. 63, No. 2, 1984; pp 290–300.
2. Goodwin, R.W.; "Meeting Clean Air Act Also Means Managing Solid Waste"; Power; Vol. 134, No. 8; August 1990; pp 55–58.
3. Resource Recovery Report; Proceedings MSW Ash Utilization Conference; October 13–14, 1988; Penn Tower Hotel; PA.
4. Poran, C.J.; Ahtchi-Ali, F.; "Properties of Solid Waste Incinerator Ash"; Journal of Geotechnical Engineering, ASCE; Vol. 115, No. 8; August 1989; pp 1118–1133.
5. Glick, H.B.; "Coal Ash Use as an Economical Cover at Sanitary Landfills"; Proceedings of the 7th International Ash Utilization Symposium; May, 1985.
6. Goodwin, R.W.; "Ash from Refuse Incineration Systems: Testing, Disposal, and Utilization Issues"; presented at MASS-APCA 33rd Technical Conference and Exhibition: Air Pollutants from Incineration and Resource Recovery; November 3–6, 1987; Atlantic City, NJ.
7. Gallagher, D.; Burde Associates; Personal Communications; 10/6/89.
8. McEnroe, B.M.; Schroeder, P.R.; "Leachate Collection in Landfills: Steady Case"; Journal of Environmental Engineering Division, ASCE; 1988, Vol. 114 (5); pp 1052–1062.
9. DeWalle, F.B., et al; "Gas Production from Solid Waste Landfills"; Journal of Environmental Engineering Division, ASCE; 1978, Vol. 104 (EE3); pp 415–432.
10. Gray, M.N.; Rock, C.A.; Pepin, R.G.; "Predicting Landfill Leachate with Biomass Boiler Ash"; Journal of Environmental Engineering Division, ASCE; 1988, Vol. 114 (2); pp 465–470.
11. Goodwin, R.W.; "Coal and Incinerator Ash in Pozzolanic Reaction Applications"; presented at MSW Ash Utilization Conference; sponsored by Resource Recovery Report; October 13–14, 1988; Tower Hotel; PA.
12. Forrester, K.; Wheelabrator Tech.; Personal Communication; 10/6/89.
13. Goodwin, R.W.; "Incinerator Ash–The Tip of the Iceberg"; presented at 8th Annual Conference on Solid Waste Management and Materials Policy; January 28–31, 1992; Sheraton New York, New York City.

FURTHER READING

EPRI CS-55269; "Utilization Potential of Advanced SO$_2$ Control By-Products"; June, 1987.
EPRI CS-5312; "Calcium Spray Dryer Waste Management: Design Guidelines"; September, 1987.
EPRI CS-6044; "Advanced SO$_2$ Control By-Product Utilization (Laboratory Evaluation)"; September, 1988.
EPRI CS-6053; "Atmospheric Fluidized-Bed Combustion Waste Management: Design Guidelines"; December, 1988.
EPRI CS-5783; "Laboratory Characterization of Advanced SO$_2$ Control By-Products: Furnace Sorbent Injection Wastes"; May, 1988.
DOE; "Waste Disposal/Utilization Study"; Contract No. DE-AC 21-88MC 25042; (1989/1990).
DOE; "Clean Coal Technology Demonstration Program, Final Programmatic Environmental Impact Statement"; No. DOE/EIS-0146; November, 1989.

CHAPTER 7

Combustion Residue Reuse Concepts and Projects

Contents

7.1. USING POWER PLANT WASTE TO SOLVE THEIR DISPOSAL PROBLEMS

Of the 131 million tons of power plant waste or coal combustion residues (CCRs), 36% are disposed of in landfills and 21% are disposed of in surface impoundments. Recent upsets from surface impoundments and landfills have created a public media furor—focusing elected and appointed officials to demand more stringent regulatory control.

During the past several months, following the Tennessee Valley Authority (TVA) ash and scrubber sludge spill in Kingston, TN, the public and media have elevated this incident—engaging elected officials and federal/state regulators. This spill occurred on December 22, 2008, when an ash dike ruptured at an 84-acre ($0.34\,km^2$) coal waste containment area at the TVA's Kingston Fossil Plant in Roane County, Tennessee. Coalfly ash slurry of 1.1 billion gallons ($4.2\,million\,m^3$) was released. It was the largest fly ash release in the United States history.

In May 2009, the U.S. Environmental Protection Agency (USEPA) signed an enforceable agreement with TVA to oversee the removal of

Combustion Ash Residue Management
http://dx.doi.org/10.1016/B978-0-12-420038-8.00007-0

coalfly ash slurry at the Kingston Fossil Fuel Plant. The cost of this cleanup is estimated at almost one billion dollars. During January 2009 Public Works Hearings, Tom Kilgore (TVA CEO) said that TVA had chosen to implement inexpensive patches instead of more extensive repairs of the holding ponds, admitting, "Obviously, that doesn't look good for us".

The New York Times advocates further regulation. "The lack of uniform regulation stems from the EPA's inaction on the issue, which it has been studying for 28 years." USEPA Administrator Lisa Jackson promised, during her confirmation hearing, to promulgate stricter coal plant waste storage regulations. Capitalizing on the political opportunity of the TVA Kingston Spill will escalate with continued congressional hearings. Elected and appointed officials will find the coal industry an easy and demonized target to impose more stringent regulations.

These incidents and consequences should not indict coal-fired power plants or the electric utility industry; unexpected costs from ten to hundreds of million dollars and public embarrassment are sufficient punishments. How to avoid such upsets should be the focus of coal-fired power plant operators and the electric utility industry.

An engineering approach reflecting demonstrated technology and recognizing CCRs' chemical and geotechnical properties, should be embraced by the electric utilities with coal-fired power plants. Commitments to regulators to develop and implement this approach would curb excessive requirements. Electric utilities should capitalize upon the industry-wide knowledge and submit to USEPA as regulatory approaches are being developed.

If increased regulation translates into using CCRs in land disposal applications, with improved methods of placement (optimal compaction) and enhanced site management (capitalizing upon concrete-like behavior of coal combustion by-products), then such stronger regulation can be justified.

Consider the following uses of power plant wastes to improve how they are land disposed:

7.2. PHYSICAL PROPERTIES OF FGD RESIDUE AND FLY ASH—RETROFITTING SURFACE IMPOUNDMENTS AS GROUT TO STRENGTHEN DIKE WALLS

The particle size of FGD residue and fly ash shows this blend could be used as a grout material to stabilize existing CCR surface impoundment dike walls. When used as a grout, the blend must be able to penetrate between the interstitial soil spaces. Grouting the existing soil dike wall would be about 90% less costly than slurry cutoff wall.

According to AECOM's 6/25/09 Summary Report, a combination of the existence of an unusual bottom layer of ash and silt, the high water content of the wet ash, the increasing height of ash, and the construction of the sloping dikes over the wet ash were among the long-evolving conditions that caused a 50-year-old coal ash storage pond breach and subsequent ash spill at TVA's Kingston Fossil Plant on December 22, 2008. Retrofit of surface impoundments, using flue gas desulfurization residue and fly ash and, where required, a cementitious additive would prevent similar dike wall failures.

7.3. RESIDUE MANAGEMENT—PLACEMENT—LANDFILL METHODOLOGY

The inherent pozzolanic-like behavior of lime-laden CCRs enables achieving improved geotechnical properties i.e. strength, permeability. Achieving liner-like permeabilities, by capitalizing upon CCR's inherent characteristics and applying proper placement control, achieves cost savings of 65% over traditional disposal methods e.g. synthetic liners.

7.4. DEMONSTRATION PROGRAM—LANDFILL AND SURFACE IMPOUNDMENT EMBANKMENTS

Considering the inherent engineering properties of CCRs justifies using this material to form surface impoundment dike walls. Approximately 27.5 million tons of CCRs are retained in surface impoundments. Preventing failure of these dike walls represents a primary issue for discussions between the electric utility industry and regulators. A demonstration program, based on the laboratory and bench-scale testing, would indicate industry willingness to address future requirements in a cost-effective manner.

The electric utility industry with their trade and research organizations are urged to commit to conducting such programs (demonstrating the application of CCRs in land disposal)—showing a 'good faith' effort to cooperate with regulatory and addresses recent coal combustion by-product disposal upsets.

7.5. BENEFICIAL USE OF COMBUSTION RESIDUE ENHANCES SITING OF POWER PLANTS

Siting combustion-based energy facilities e.g. waste-to-energy/resource recovery, coal-fired power plants demand gaining public approval. These facilities are subject to the NIMBY or Not-In-My-Backyard syndrome.

Pollution control technology and advanced management techniques achieve compliance with environmental regulations. Such regulatory acceptance, often, fails to convince the local citizenry. Too often coal-fired power plants have been maligned by environmental extremists. They confuse the public, media, and elected/appointed officials. Although the facility will create some power plant jobs, enlarge the local tax base and reduce local electricity cost, these benefits are not tangible to most of the community. Frequently, they claim that most of the energy will be sent outside the area for the benefit of the power company; the local populace wants to know "what's in it [the power plant] for me". Integrating large-scale ash utilization projects enhances benefits to host communities by promoting development of abandoned property. Such local development offers tangible benefits e.g. more jobs and commercial opportunities.

7.6. BENEFICIAL USE OF DREDGED SOLIDS—STRUCTURAL FILL BROWNFIELDS

Ash stabilized and solidified dredged solids for the resultant material use as structural fill to raise elevation of a 15-acre Brownfields Site by 8 ft. The stabilized contaminated dredgings (50,000 CY) were used as a cap and structural fill enabling harbor deepening and waterfront development.

A stabilization/solidification (S/S) concept was developed to treat hazardous dredged solids, depicted by 7.1.

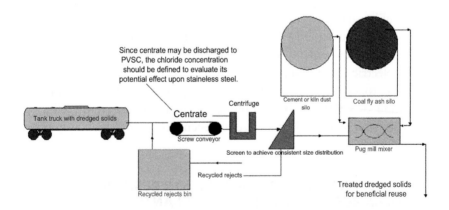

Figure 7.1 *Schematic Treatment Dredged Solids-Yielding Structural Fill Material.*

The S/S utilized ash as an additive enabling the treated material to be used as structural fill to raise elevation of a 15-acre Brownfields Site by 8 ft. Stabilized contaminated dredgings were used as a cap and structural fill enabling harbor deepening and waterfront development.

The S/S process proved the most likely scenario, since its resultant product could achieve a soil-like consistency (for use as a fill or base material) and concrete-like characteristics (for use as a final cap). Appreciating the relationship between ash's pozzolanic ingredient and desired geotechnical properties facilitated attaining such dual usage. The addition of ash with Portland cement (or lime kiln dust) induces pozzolanic behavior achieving permeability $\approx 10^{-7}$ cm/s with concrete-like compressive strengths and bearing capacities. Just adding the fly ash achieves a soil-like final product characteristic.

Remediation and demolition of 20 abandoned buildings occurred on 15 acres of previous chemical process manufacturing facility. Besides removed asbestos and lead-based paint prior to buildings' demolition, a variety of hazardous/toxic materials removal/disposal including PCB's, organic solvents, fatty acids, etc. This waterfront site required dredging of its dock. The dredged material contained dioxins, heavy metal, and organic contaminants (e.g. PAH's).

7.6.1. Regulatory Considerations

Discussions and negotiations with regulator officials evolved a "Dirty to Dirty" working concept. Given that the site was contaminated but no more severely than the proximate region and that the treated dredged solids' environmental characterizations was no worse than deposition site, use of dredged material as a structural fill would appear compatible. Coordination required between remediation activities and use of dredged material— ensure compatibility. Applying solidification stabilization to dredged solids must coordinate with regulators to determine compatibility. The characteristics of treated material must conform to final site characteristics.

To minimize liability: marry final site determination under regulatory remediation requirements to the concept of using dredged material as cap with deed restriction. Technically if dredged solids are no worse than land on which it will be placed then a null effect occurs. Liability still exists with treatment and placement. The application of optimal placement of the treated dredged material mitigated this liability. Identifying the source of dredged material required coordination with the Army Corps of Engineers and the New York/New Jersey Port Authority.

7.6.2. Economic Considerations

The engineering process of stabilizing dredged solids costs $11.50/ton vs traditional fill costing $9–11/CY. The negotiated tipping fee of $5/CY achieved project economic justification. The beneficial use dredging schedule includes: demonstration phase = 50,000 CY; project phase I = 110,000 CY; and project phase II: 3×10^6 CY.

Capping a Brownfield site with stabilized dredged material represents stronger economic justification. Using bentonite clay as a cap costs $160/CY. Environmental cleanup and remediation of contaminated dredged material—costs associated with removing and treating—reach $146/CY. Capping with stabilized dredged material yields a 90+% savings.

Paying the electric utility a nominal price for their ash would reduce disposal costs while maintaining considerable remediation costs savings.

7.7. BENEFICIAL USE OF COMBUSTION RESIDUE—MINE RECLAMATION

Combustion residue was used as a grout material to stabilize an abandoned underground roof and pillar mine as depicted by Figure 7.2.

A 50–60% (by weight) water/fly ash slurry was gravity fed to a depth of about 150 ft, filling the spaces between the roof and deteriorated pillars. Backfilling of dry ash achieved final closure of each roof and pillar segment. Collected runoff from this slumped fly ash grout was collected, recycled as makeup water and as hydration water to react with the dry backfilled ash. Mine operation water management practice recycles over 90% of the incoming water. Typically, lime-laden fly ash reflects pozzolanic characteristics of enhanced strength, lower permeability, and heavy metal encapsulation. Cementitious fly ash (type "F" and "C") were used to attain minimum UCS of 40 PSI. Although a UCS of 40 PSI represented the design basis, subsequent field testing showed attaining UCS exceeded 100 PSI. More than 50,000 tons/year of type "F and C" fly and bottom ash will be used to prevent pillar deterioration for a 7–10 year stabilization period.

Bottom ash, dredged from ponds, behaves as a granular soil, exhibiting compressibility and permeability. Regulatory required tests, including Toxicity Characteristic Leachate Procedure, show conformity to criteria—confirming the nonhazardous characteristics of these residues. Bottom ash serves as a drainage blanket; facilitating sloughing water from fly ash slurry. Approximately one-half, of the nonrecycled water, passes through the

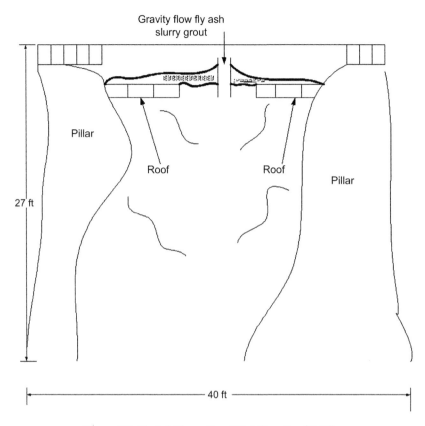

Figure 7.2 *Fly Ash Slurry Grout Stabilizes Roof & Pillar.*

bottom ash prior to its permitted discharge to the local surface water. This intermittent flow's contribution to the local surface water was deemed minimal with respect to quantity and quality. The underground mine has been redeveloped for commercial activities.

7.7.1. Economic Considerations

The underground mine development plans for over 1 million ft² of space available for commercialization. Upon stabilization, the underground mine will maintain a constant temperature of 57 °F. Initial job creation estimated that about 150 people would be employed in the commercial underground facilities. Later projections increased to claim that 500–1700 new jobs would be created.

The coal-fired power plant spent about $500,000/year to transport their ash to the mine reclamation site. At a transport cost of $10/ton, the electric utility saved about 50% of avoided disposal cost. Based on typical commercial grout costs of $50–60/ton, the developer saved several millions of dollars by using the fly ash (costs are in $1998).

Commercialization of the stabilized mine exhibits several "Economic Benefits". Due to the 57 °F constant temperature, an 80–90% energy reduction was achieved. The preexisting room (roof and pillar configuration) results in a development cost savings of $0.60/SF–$1.20/SF. The saved structural costs realize a 40–50% reduction over traditional surface development cost. The underground mine accommodates a 30 day turnaround for "quick move" (i.e. ready occupancy) (costs are in $1998).

Local elected officials supported this project. Rigorous engineering studies and investigations convinced regulatory officials of this projects engineering and environmental viability.

FURTHER READING

Dewan, S.; "Hundreds of Coal Ash Dumps Lack Regulation"; The New York Times; January 7, 2009.

Goodwin, R.W.; Combustion Ash/Residue Management – An Engineering Perspective; Noyes Publications/William Andrew Publishing; Mill Road, Park Ridge, NJ; 1993 (ISBN: 0-8155-1328-3) (Library of Congress Catalog Card No.: 92-47240).

RS Means; Building Design Construction Data; Kingston MA; 2006.

CHAPTER 8

Coal Combustion Residue Disposal Options

Contents

8.1. PREFACE—ENGINEERING APPROACH

Coal combustion residues (CCRs) are presently regulated as solid waste (Subtitle D) under the Resource Conservation Recovery Act. Such classification promotes beneficial use by end-users i.e. mitigating excessive liability. According to the US Environmental Protection agency (USEPA), about 131 million tons of coal combustion residuals—including 71 million tons of fly ash, 20 million tons of bottom ash and boiler slag, and 40 million tons of flue gas desulfurization (FGD) material—were generated in the US in 2007. Of this, approximately 36% was disposed of in landfills, 21% was disposed of in surface impoundments, 38% was beneficially reused, and 5% was used as minefill. Stringent regulation, as Subtitle C (hazardous waste), would impose a perceived liability upon end-users; greatly reducing beneficial use

Combustion Ash Residue Management
http://dx.doi.org/10.1016/B978-0-12-420038-8.00008-2

opportunities. Mandatory use of synthetic liners—would not have prevented dike wall failure and fails to consider inherent engineering characteristics of CCRs.

If increased regulation translates into improved methods of placement (optimal compaction), better monitoring (via monitoring wells with periodic reporting), and enhanced site management (capitalizing upon concrete-like behavior of coal combustion by-products), then such stronger regulation can be justified. But, if greater regulatory scrutiny imposes a bureaucratic burden on operating coal-fired power plants without understanding characteristics of these by-products, future spills, leaks, and dike failures will continue. For example, should USEPA mandate synthetic liners—a costly approach—the inherent behavior of coal combustion products to achieve liner-like permeability between 10^{-5} and 10^{-7} cm/s will have been ignored (1).

8.2. COMPOSITION—POZZOLANIC CHEMISTRY

An appreciation of the chemical composition of fly ash and FGD residuals provides insight for capitalizing upon their inherent pozzolanic behavior. Lime (CaO) in the presence of silica (SiO_2), alumina (Al_2O_3) and calcium sulfate ($CaSO_4$) form sulfo-alumina hydrates (ettringites) and calcium silica hydrate (tobermorite), as represented by the following:

$$\text{Ettringite: } 3CaO \cdot Al_2O_3 \cdot 3CaSO_4 \cdot 28\text{–}32H_2O$$

$$\text{Tobermorite: } CaO \cdot SiO_2 \cdot nH_2O$$

8.2.1. Fly Ash—Chemical Composition

Particulate control devices (e.g. electrostatic precipitators, baghouses) remove fly ash particulates from flue gas for subsequent collection and beneficial use and/or disposal. The following tabulation shows the significant percent components (percent by weight) of CaO, SiO_2, and Al_2O_3.

Component	Bituminous	Subbituminous	Lignite
SiO_2 (%)	20–60	40–60	15–45
Al_2O_3 (%)	5–35	20–30	20–25
Fe_2O_3 (%)	10–40	4–10	4–15
CaO (%)	1–12	5–30	15–40

8.2.2. FGD—Chemical Composition

SO_2 is an acid gas and thus the typical sorbent slurries or other materials used to remove the SO_2 from the flue gases are alkaline. The reaction-taking place in wet scrubbing using a $CaCO_3$ (limestone) slurry produces $CaSO_3$ (calcium sulfite). When FGD were first introduced this so called "FGD sludge" was ponded. But some FGD systems go a step further and oxidize the $CaSO_3$ to produce marketable $CaSO_4 \cdot 2H_2O$ or gypsum.

8.2.3. Air Pollution FGD Chemistry

The following chemical reactions depict the formation of final end-products containing $CaSO_3 \cdot \frac{1}{2}H_2O$ and $CaSO_4 \cdot 2H_2O$

$$SO_2 + CaCO_3 + \tfrac{1}{2}H_2O \rightarrow CaSO_3 \cdot \tfrac{1}{2}H_2O$$

Some FGD systems employ "forced oxidation" to convert the $CaSO_3$ (calcium sulfite) to produce marketable $CaSO_4 \cdot 2H_2O$ (gypsum):

$$SO_2 + \tfrac{1}{2}O_2 + CaCO_3 + 2H_2O \rightarrow CaSO_4 \cdot 2H_2O + CO_2$$

8.3. BLENDING OF FGD RESIDUE WITH FLY ASH—USE AS LINER AND EMBANKMENT

The following tabulation shows the final composition of $CaSO_3$, $CaSO_4$, and fly ash. When blended together (i.e. in a pug mill) the resultant material can be landfilled to achieve in-situ pozzolanic reactants and behavior.

Major Components of FGD Scrubber Material and Fly Ash from Different Coal Types and Scrubbing Processes (Percent by Weight)

Type Coal	% Sulfur	FGD Process	CaSO₃	CaSO₄	CaCO₃	Fly Ash			
						SiO₂	Al₂O₃	Fe₂O₃	CaO
Bituminous	3.5	Lime	72	4	1.5	9.6	4.8	6	1.6
Bituminous	2.9	Limestone	21	23.5	23	12.6	6.3	7.9	2.1
Bituminous	2.5	Limestone—forced oxidation							
	1.5		58.5	3.5	14	7	8.8	2.3	–
Subbituminous	0.75	Limestone	10	20	30	20	10	2.8	7

A relationship between mineralogical composition and strength has been developed (1). The sum of "$SiO_2 + Al_2O_3 + Fe_2O_3$ divided by CaO" varies linearly with "unconfined compressive strength (UCS)"; where constituents are expressed as % by weight and UCS expressed as thousands of psi. Direct linear extrapolation predicts UCS ranging from 6500–17,900 psi; however, extrapolation exceeded data points by order-of-magnitude. Applying highly conservative order-of-magnitude reduction yields reduced UCS predictions ranging from 650 to 1790 psi. These raw FGD sludge and fly ash blends adjusted UCS values support using this resultant material as dike embankment. Adding Portland cement would increase strength—supporting such load-bearing application and justifying laboratory—demonstration studies.

8.4. GEOTECHNICAL PROPERTIES

Geotechnical Properties of Typical Calcium Sulfite FGD Scrubber Material

Geotechnical Property	Dewatered	Stabilized	Fixated
Shear strength—internal friction angle	20°	35°–45°	35°–45°
Permeability (cm/s)	10^{-4} to 10^{-5}	10^{-6} to 10^{-7}	10^{-6} to 10^{-8}
28-Day unconfined compressive strength	–	170–340	340–1380
(kPa) (lb/in^2)	–	25–50	50–200

The geotechnical properties listed above represented dewatered FGD sludge (i.e. vacuum filter, centrifuge), stabilized FGD sludge (i.e. blended with fly ash), and fixated FGD sludge (addition of CaO or Portland cement to fly ash—sludge blend). Synthetic liner-like permeabilities of 10^{-7} cm/s or less can be attained by (1) blending ash with FGD or (2) adding Portland cement or CaO to the blend. Based on in-situ field results, when 6.6–10% Portland cement was added to blend, permeabilities from 10^{-7} to 10^{-9} cm/s were attained (1). Since the internal angle of friction (Ø) ranged from 35° to 45° for the blend with or without Portland cement or CaO, this material should behave like a cohesive soil and could be used as an embankment material. For instance, addition of Portland cement or CaO could form a stable dike wall for CCBs surface impoundment (2).

8.5. PHYSICAL PROPERTIES OF FGD RESIDUE AND FLY ASH—RETROFITTING SURFACE IMPOUNDMENTS

The particle size of FGD residue and fly ash shows this blend could be used as a grout material to stabilize existing CCR surface impoundment dike walls. When used as a grout, the blend must be able to penetrate between the interstitial soil spaces.

8.5.1. Fly Ash—Particle Size

To be used in cement or concrete applications (i.e. grout), fly ash should conform to ASTM C618—either as Class C or F—depending on their chemical composition. Seventy five percent of the ash must have a fineness of 45 μm or less, and have a carbon content, measured by the loss on ignition (LOI), of less than 4%. In the US, LOI needs to be under 6%. Since not all fly ashes meet ASTM C618 requirements, this makes it necessary that fly ash used in concrete needs to be processed using separation equipment like mechanical air classifiers or similar separation equipment.

8.5.2. FGD Residue—Particle Size

The tabulation shown below indicates that FGD residue most closely resembles a silt-like soil, having a fine-grained consistency of less than 0.074 mm size (3).

Typical Particle Sizes of FGD Scrubber Material

Property	(Unoxidized) Calcium Sulfite	(Oxidized) Calcium Sulfate
Particle Sizing (%)		
Sand size	1.3	16.5
Silt size	90.2	81.3
Clay size	8.5	2.2
Specific gravity	2.57	2.36

8.6. FLY ASH AND FGD RESIDUE BLEND—PARTICLE SIZE—SUITABLE FOR GROUTING RETROFIT

Adding fly ash and FGD residue would yield a final material whose size would be less than 0.074 mm—suitable as a grout material even with fine-grained Portland cement addition. Grouts formed from fly ash and FGD residue have been demonstrated as grout material in Maryland (2). CCR retention ponds are similar to earthen dams; strengthening dike walls of these dams often employ slurry cutoff walls. Grouting existing soil dike wall would be about 90% less costly then slurry cutoff wall (4).

8.7. RESIDUE MANAGEMENT—PLACEMENT—LANDFILL METHODOLOGY

The inherent pozzolanic-like behavior of lime-laden CCRs enables achieving improved geotechnical properties i.e. strength, permeability. Sound engineering practice of CCR placement recognizes the relationship between achieving desired geotechnical property and optimal moisture content while maintaining adequate "water of solubilization" (or % final solids) to ensure reacting pozzolanic constituents with free available lime adhering to these principles requires proper placement control management—field compaction and water addition. Figure 8.1 depicts this methodology. Improved field placement (i.e. compaction, addition of dust suppression water) of CCR could increase density and reduce permeability to decrease leachate rate through the buried CCR. Achieving liner-like permeabilities, by capitalizing upon CCR's inherent characteristics and applying proper placement control, achieves cost savings of 65% over traditional disposal methods, e.g. synthetic liners (1). USA regulatory officials should consider incorporating these principles into residue management recommendations. Recognition and implementation of these principles would confirm that CCRs can be properly managed—to alleviate concerns—providing a cost-effective approach to future regulatory control.

Achieving liner-like permeabilities represent demonstrated technologies. Achieving low permeabilities and enhanced compressive or bearing strength are known and recognized methodologies. Using CCRs as embankment material requires additional demonstration.

8.8. DEMONSTRATION PROGRAM—LANDFILL AND SURFACE IMPOUNDMENT EMBANKMENTS

Considering the inherent engineering properties of CCRs justifies using this material to form surface impoundment dike walls. Approximately 27.5 million tons of CCRs are retained in surface impoundments. Preventing failure of these dike walls represents a primary issue for

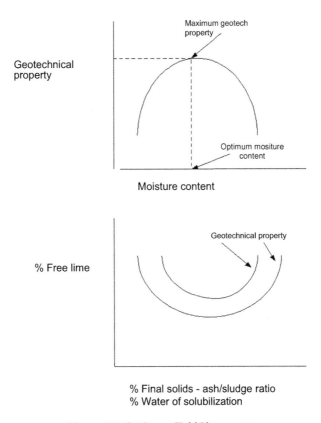

Figure 8.1 *Optimum Field Placement.*

discussions between the electric utility industry and regulators. A demonstration program, based on laboratory and bench-scale testing, would indicate industry willingness to address future requirements in a cost-effective manner.

Laboratory and bench-studies could include triaxial testing, determination of angle of internal friction, and slope stability analyzes. Additive studies could include varying percentages of CaO and Portland cement. These studies would also investigate using FGD and fly ash and grout material—with and without additives.

Demonstration programs could be conducted for new landfill and surface impoundment embankment stability. Demonstration programs should be conducted for using fly and FGD sludge as retrofit grout material to strengthen existing embankments (landfill) and dike wall (surface impoundments). Based upon their inherent characteristics, suitable engineering properties should be attained at cost-effective results.

The electric utility industry with their trade and research organizations are urged to commit to conducting such programs—showing a "good faith" effort to cooperate with the regulatory and addresses recent CCR disposal upsets.

8.9. STAFF TRAINING

During execution of demonstration programs, operating staff should participate to gain field knowledge regarding proper land disposal management. The use of in-situ methods to optimize compaction and percent water of solubilization would be learned in the field. Encountering and adjusting for unexpected geotechnical behavior also enhances staff learning.

Besides learning from demonstration programs, upsets from CCR land disposal projects offer additional "lessons learned" opportunity. For example, an ash monofill experienced leachate piping backup and overflow. Lime was added to the ash to immobilize lead and cadmium. This lime/ash material prematurely reacted to form a concrete-like pozzolanic material that reduced leachate collector pipe opening—causing a backup and overflow. This upset condition, causing excessive groundwater levels, could have avoided by applying the publicized, peer-reviewed, engineering placement principles and methodology. Operators of CCR landfill facilities are urged to apply this methodology (5).

8.10. DISCUSSION

An engineering approach, reflecting demonstrated technology and recognizing coal combustion products' chemical and geotechnical properties, should be embraced by the electric utilities with coal-fired power plants. Commitment to develop and implement this approach would curb excessive regulatory requirements and allay public concerns. Electric utilities should capitalize upon the industry-wide knowledge and submit to USEPA as regulatory approaches are being developed. Application of CCRs' engineering characteristics in land disposal projects would provide a cost-effective approach to pending regulatory negotiations. The electric utility industry and their trade-research organizations are urged to consider this engineering perspective in dealing with governmental agencies, elected and appointed officials, and public media groups.

REFERENCES

1. Goodwin, R.W.; "Combustion Ash/Residue Management – An Engineering Perspective"; Noyes Publications/William Andrew Publishing; Mill Road, Park Ridge NJ; 1993 (ISBN: 0-8155-1328-3); (Library of Congress Catalog Card No.: 92-47240).
2. Wattenbach, H.L.; "An Evaluation of Free-Lime Containing By-products to Produce CCB Grouts for Use in AMD Abatement"; 1999 International Ash Utilization Symposium; University of Kentucky, Paper No. 20.
3. Zilly, R.G. (editors); Handbook of Environmental Civil Engineering; Van Nostrand Reinhold Co., New York; 1975.
4. RS Means; "Building Design Construction Data"; Kingston MA; 2006.
5. Goodwin, R.W.; "Avoiding Ash Landfill Operating Mistakes"; Energy Pulse.Net; March 27, 2003.

FURTHER READING

http://www.netl.doe.gov/technologies/coalpower/cctc/topicalreports/pdfs/Topical24.pdf www.flyash.info/2009/189-fitzgerald2009.pdf.

Rusch, K.A.; "Development of CCB Fill Materials for Use as Mechanically Stabilized Marine Structures"; CBRC Project Number: CBRCM11; Subcontract No. 98-166-LSU; August 30, 2002.

Science Applications International Corporation; "Technical Background Document on the Efficiency and Effectiveness of CKD Landfill Design Elements; EPA Contract 68-W4-0030, draft July 18, 1997.

USDOE; "Clean Coal Technology – Coal Utilization By-Products"; Technical Report No. 24; August 2006.

Berger, E.; Fitzgerald, H.B.; "Use of Calcium-Based Products to Stabilize Ponded Coal Ash Techniques and Results"; 2009 World of Coal Ash, Lexington, KY, May 4–7, 2009.

CHAPTER *9*

Lessons and Outlook

Contents

9.1. PREFACE—SETTLING COAL ASH CONTROVERSY

The EPA's May 4, 2010 proposal to regulate coal combustion residues (CCRs) as either a Class C hazardous waste or a Class D solid waste is a political Chinese Menu; although regulating coal combustion by-products under either option, the proposal fails to provide regulatory direction. For a material to be classified as hazardous, failure of meeting the Toxicity Characteristic Leachate Procedure (TCLP) must occur. Those, who arbitrarily denote CCRs as hazardous or toxic, do so without substantive TCLP data. Retaining the Bevill exemption, allowing for beneficial use of CCRs, reflects the industry practice of deploying about 40% of the residue stream. Nonetheless should EPA decide to classify CCRs as hazardous or Class C—the potential for end-user liability could reduce by-product utilization.

I advocate the engineering approach of recognizing the inherent properties of CCRs when land disposal is considered. Referencing the EPA

Combustion Ash Residue Management
http://dx.doi.org/10.1016/B978-0-12-420038-8.00009-4

public record of electric utilities, cited for potential ponding concerns, the introduction of 68% paste (consisting of flue gas desulfurization (FGD) slurry) ash-formed dike walls should contribute to improved permeability and shear strength. Stabilized FGD sludge (i.e. blended with fly ash) has been used as a grout material. Synthetic liner-like permeabilities of 10^{-7} cm/s or less can be attained by blending ash with FGD sludge. Based on geotechnical testing, this blend should behave like a cohesive soil and could be used as embankment material.

9.2. GOOD ENGINEERING PRACTICE COULD HAVE AVOIDED THE TENNESSEE VALLEY AUTHORITY INCIDENT

CCRs (including fly ash, bottom ash and FGD sludge) is regulated as a solid waste (i.e. nonhazardous). About 30% of these wastes are utilized in road construction, bricks, etc. (fly ash) and cement wallboard production (FGD by-product gypsum). Rather than pond and landfill these materials, Tennessee Valley Authority (TVA) could have opted to investigate their beneficial reuse. Several electric utilities dedicate staff to develop markets and end-users for their CCRs/by-products (e.g. TECO, AEP). The failure of the pond dike walls could have been avoided if TVA had appreciated the well-documented behavior of ash with FGD waste forming a pozzolanic material suitable as a dike wall instead of using earthen material.

Instead of creating unnecessary legislation or investigate undemonstrated technology, the application of Good Engineering Practice (based on over 20 years of field operating experience) should be part of a coal-fired power plant's management philosophy.

The electric utility sector has altered its position in the wake of TVA's misfortune. The industry now agrees that federal rules are necessary—but only those that would supplement current coal ash disposal laws on the states' books. Regulating the by-product as a hazardous waste is an ill-advised, power companies reason, maintaining that the shift in public policy would result in less recycling and more pollution. Industry's conciliatory position may ultimately appease the EPA. But activists are unconvinced and are trying to persuade the agency that the spill could have been avoided with tougher national laws.

If increased regulation translates into improved methods of placement (optimal compaction), better monitoring (via monitoring wells with periodic reporting), and enhanced site management (capitalizing upon concrete-like behavior of CCRs), then I support stronger regulation. But, if greater regulatory scrutiny imposes a bureaucratic burden on operating coal-fired power plants without understanding characteristics of these residues, future spills, leaks, and dike failures will continue. For example, should U.S. Environmental Protection Agency (USEPA) mandate synthetic liners—a costly approach—the inherent behavior of CCRs to achieve liner-like permeability between 10^{-5} and 10^{-7} cm/s will have been ignored.

Electric utilities should capitalize upon the industry-wide knowledge and submit to USEPA as regulatory approaches are being developed.

Rather than create unnecessary legislation or investigate undemonstrated technology, the application of Good Engineering Practice (based on over 20 years of field operating experience) should be part of a coal-fired power plant's management philosophy.

9.2.1. TVA Failed to Apply Good Engineering Practice

TVA ignored geotechnical engineering consultants' recommendations to strengthen dike walls (1). Since May 2009, the EPA has been conducting on-site assessments of coal residue impoundments and pondings at electric utilities. The EPA on Thursday, February 4, 2010, released action plans developed by 22 electric utility facilities with coal residue impoundments, describing the measures the facilities are taking to make their impoundments safer. The action plans address recommendations from assessments of 43 impoundments and many electric utilities have begun implementing the recommendations. For instance, one electric utility used a grout consisting of FGD sludge and fly ash to strengthen a dike wall; USEPA and state regulators accepted this methodology—demonstrating the beneficial use of CCRs (2). Many of the electric utilities hired consultants to assist in developing their action plans. Now that these plans have been implemented, the issue of future regulations impact proposed coal-fired plants and those in construction—seeking applicable permits.

Land-filling CCRs offer a more manageable site option to achieve optimal geotechnical properties, i.e. achieve liner-like permeabilities by capitalizing on the material's inherent pozzolanic properties. Bottom ash ponding requires a more rigorous geotechnical design and monitoring program. Beneficial use of FGD by-product gypsum and fly ash for building materials are ongoing commercial successes.

Regulating coal residue as hazardous based on the Kingston ash spill, in my opinion, is not justified. The cause of the spill was due to TVA ignoring their geotechnical consultants concern regarding the structural stability of the pond's dike wall. Over the past year, the USEPA has investigated about 50 CCR impoundments—making recommendations to respective electric utilities retrofitting where required. Thus, a database of engineering criteria exists to develop mandatory guidelines for future impoundments—preventing another dike wall failure. The hazardous waste designation would greatly inhibit the beneficial use of CCRs, e.g. FGD by-product gypsum and fly ash used as additives for construction material—due to the potential risk of litigation liability by end-users.

The Kingston ash spill occurred because engineering judgment was ignored—applying an engineering approach to future facilities would not only avoid another accident but continue to use beneficial coal combustion by-products based on demonstrated engineering applications.

9.3. RETROFIT COSTS

The USEPA possible classification of CCRs as Section C—hazardous waste introduces expensive retrofit costs to coal-fired power plants. While the EPA estimated that additional disposal expenses would hover around $1 billion a year under a "hazardous" designation for coal ash, the utility industry argued that the figure was "significantly understated". An economic analysis commissioned by the Edison Electric Institute and the Utility Solid Waste Activities Group, two associations of power companies, pegged the costs as high as $13 billion a year. On October 16 another draft regulation was submitted to the Office of Management and Budget—details and associated costs are yet to be made public.

9.3.1. How to Avoid Such Upsets Should be the Focus of Coal-fired Power Plant Operators and the Electric Utility Industry

An engineering approach, reflecting demonstrated technology and recognizing CCRs' chemical and geotechnical properties, should be embraced by the electric utilities with coal-fired power plants. Commitments—to regulators—to develop and implement this approach should curb excessive requirements. Electric utilities should capitalize upon the industry-wide knowledge and submit to USEPA as regulatory approaches are being developed.

9.4. REGULATORY AGENCIES SHOULD EMPHASIZE SOUND ENGINEERING PRACTICE

Recent Congressional legislation and USEPA proposed regulations would modify how CCRs are managed and controlled.

9.4.1. The Coal Residuals Reuse and Management Act of 2013

Rep. David McKinley's (R–W.Va.), The Coal Residuals Reuse and Management Act of 2013 (H.R. 2218) retains some of the same principles the lawmaker proposed in a similar legislation that passed the House by a bipartisan vote of 267 to 144 in 2011. It also contains requirements for groundwater monitoring at all structures that receive coal ash after the legislation's enactment and corrective action for unlined, leaking impoundments within a specified time period, as outlined in a previous Senate bill. This adopted legislation also includes new provisions to ensure structural stability, including a consultation with state dam safety officials, a periodic evaluation to identify structural weakness and potentially hazardous conditions, and the creation of an emergency action plan for high-hazard structures.

I advocate application of sound engineering practice based on concrete chemistry and civil geotechnical engineering. The upset condition (described in Chapter 5), causing excessive groundwater levels, could have been avoided by applying the publicized, peer-reviewed, engineering placement principles and methodology. Operators of ash landfill facilities are urged to apply this methodology.

When implementing beneficial use applications these same principles and techniques apply. For those utilization modes requiring a soil-like consistency (e.g. structural fill, sanitary landfill cover, road construction base course), attaining optimum moisture content and maximum density achieves desired geotechnical properties. When a concrete-like, pozzolanic reaction is desired (e.g. remediation cap, stabilization/solidification of contaminated material), the optimum percent water of solubilization should be introduced. Solubilizing the pozzolanic constituents capitalizes upon the ashes' mineralogy or supplements the ash with nonproprietary additive. Addition of "water of solubilization" also mitigates fugitive dusting.

USA regulators should follow Environment Canada's lead by also recommending this engineering placement methodology. Good design cannot compensate for human operator error—proper training and monitoring recommendation provided.

9.4.2. USEPA Control of CCP (Coal Combustion Products) Discharges to Surface Waters

On April 20, 2013 the EPA proposed regulating the water discharge from steam-electric power plants primarily coal and nuclear facilities, which account for half the toxic pollutants in U.S. water bodies. Regulators proposed a range of options. Costs to comply with the rules range from $185 million to $954 million, agency economists estimated. The more expensive options would target more units and more sources of waste, resulting in greater pollution reduction, it said (3).

9.4.2.1. Holding Ponds

Because some coal-fired power plants mix the ash with water and store it in ponds, the effluent limitations will affect how the agency regulates coal ash. According to the EPA, its proposal curbing water discharges from power plants would prod many utilities to switch to the so-called dry handling of coal ash, thus curbing or eliminating the use of the ash ponds. Those ash ponds can rupture or leak.

The EPA had proposed two approaches for such rules in 2010, including labeling it hazardous, which would add stricter standards for plant owners. Republicans in Congress and groups lobbying for companies such as Headwaters pressed the agency to abandon that approach.

The new water-discharges regulations will add to measures on coal-fired generators issued since President Barrack Obama took office in 2009. The administration's most-expensive rule would cut mercury and other toxic air emissions from such plants. The April 20 measure would prevent similar minerals, arsenic and mercury, in water releases from the plants.

Politics aside, sound engineering judgment supports the nonhazardous classification of CCRs—based upon decades of operating experience.

9.5. ECONOMIC SUMMARY COMMENTS

The USEPA has proposed two classifications for these materials: hazardous or nonhazardous. Beneficial use of ash and FGD sludge (i.e. FGD by-product gypsum) is exempted; the former used as additive to construction materials and the latter used in wallboard and cement production (e.g. Tampa and Seminole Electrics). Since approximately 40% of these wastes are used, if USEPA opts for hazardous waste classification—beneficial use exemption notwithstanding—the threat of potential litigation would defer end-users (e.g. LaFarge Cement, US Gypsum) from reuse.

If the by-product would be regulated as a hazardous material, that would cost industry $1.5 billion a year whereas if it is viewed as a nonhazardous material, it would run $600 million a year. This would result in higher construction material prices and increase electric utilities disposal costs and electricity generation rates.

Dry land-filling CCRs offer a more manageable site option to achieve optimal geotechnical properties, i.e. achieve liner-like permeabilities by capitalizing on the material's inherent pozzolanic properties. Ponding of bottom ash, as discussed previously, requires a more rigorous geotechnical design and monitoring program. Beneficial use of FGD by-product gypsum and fly ash for building materials are ongoing commercial successes.

9.5.1. FGD Slurry Geotechnical Properties—As Dike Embankment Material

Introducing FGD slurry as a retrofit grout material achieved 35°–45° angle of internal friction—behaving like cohesive soil suitable as an embankment material and liner-like permeabilities $\leq 10^{-5}$ cm/s—liner like. As an additional option, combine fly ash and FGD slurry with particle size <0.074 mm with Portland cement to achieve groutable consistency. Such use of CCRs as retrofit grout to strengthen dike walls should result in cost savings. Using FGD slurry (combined with fly ash and/or with Portland cement) would be 90% less expensive than slurry cutoff wall. In comparison consider the average unit cost converting bottom ash handling system ~$20 million ($10–40 million) and the average unit cost converting fly ash handling system ~$10–15 million ($20–80 million) (4).

Sound engineering judgment supports the nonhazardous classification of CCRs—based upon decades of operating experience. Consider using CCRs as a retrofit material, i.e. applying a cost-effective, engineering solution.

REFERENCES

1. Goodwin, R.W.; "Land Disposal of Coal Combustion Byproducts—Upsets and Implications (Part 1)"; Energy Pulse Weekly; October 6, 2009.
2. Goodwin, R.W.; "Management of Coal Combustion Products (CCP's)—Avoiding Disposal And Utilization Upsets"; EPA Rule Making And Coal Combustion Products; Electric Utility Consultants Inc.; March 14–15, 2011; Denver, Colorado.
3. Drajem, M.; "Coal-Ash Recyclers Seen Aided in EPA's Water-Discharge Plan"; Bloomberg News; April 22, 2013.
4. Morris, L.; "Coal Ash Handling Rules Changes Generation Face"; Power, February 2011; pp 42–44.

FURTHER READING

Goodwin, R.W.; "Land Disposal: Fly Ash and Sludge"; presented at American Power Conference, Chicago, Illinois, April, 1980.

Goodwin, R.W.; "Resource Recovery from Flue Gas Desulfurization Systems"; Journal of the Air Pollution Control Association; Vol. 32, No. 9, September 1982; pp 986–989.

Goodwin, R.W.; "Rational Approach for Land Disposal of Residues from Thermal Combustion"; presented at the APCA International Specialty Conference on Technical Issues and Problems of Hazardous Waste Permitting; Orlando, FL; March 2–5, 1986.

Goodwin, R.W.; "Pozzolanic Behavior of Ash in Terms of Landfill and Utilization Management"; presented at Northwest Center for Professional Education Conference – "MSW Incinerator Ash"; December 1–2, 1988; Raddison Plaza Hotel, Orlando, FL.

Goodwin, R.W.; "Combustion Ash/Residue Management – An Engineering Perspective"; Noyes Publications/William Andrew Publishing; Mill Road, Park Ridge NJ, 1993; (ISBN: 0-8155-1328-3) (Library of Congress Catalog Card No.: 92-47240).

Goodwin, R.W.; "Beneficial Use of Ash Enhances Siting of Coal-Fired Power Plants"; Energy Pulse; October 22, 2002.

Goodwin, R.W.; "How Use of Coal Combustion Products Overcome Nimby Siting Issues"; Proceedings of American Coal Ash Association 15th Int'l symposium on Management & Use of Coal Combustion Products; January 27–30, 2003; St. Petersburg FL.

INDEX

Note: Page numbers with "f" denote figures; "t" tables.

Printed and bound by CPI Group (UK) Ltd, Croydon, CR0 4YY

08/05/2025

01864907-0004